Shock Wave Dynamics

Derivatives and Related Topics

Shock Wave Dynamics

Derivatives and Related Topics

GEORGE EMANUEL

CRC Press
Taylor & Francis Group
Boca Raton London New York

CRC Press is an imprint of the
Taylor & Francis Group, an **informa** business

CRC Press
Taylor & Francis Group
6000 Broken Sound Parkway NW, Suite 300
Boca Raton, FL 33487-2742

First issued in paperback 2019

ISBN-13: 978-1-4665-6420-6 (hbk)
ISBN-13: 978-0-367-38047-2 (pbk)

Library of Congress Cataloging-in-Publication Data

Emanuel, George.
    Shock wave dynamics : derivatives and related topics / George Emanuel.
        p. cm.
    Includes bibliographical references and index.
    ISBN 978-1-4665-6420-6 (hardback)
    1. Shock waves--Mathematical models. 2. Derivatives (Mathematics) I. Title.

QC168.85.S45E545 2012
531'.1133--dc23   2012028837

Visit the Taylor & Francis Web site at
http://www.taylorandfrancis.com

and the CRC Press Web site at
http://www.crcpress.com

Contents

Preface

An analytical approach is used to establish structural properties of curved shock waves. Special care is given to assumptions, implementation requirements, and the frequent use of illustrative examples. These provide partial verification of the preceding analysis. This book is a research monograph and textbook with problems for which a Solutions Manual is available. The text is at the graduate level, in part, because some of the mathematical techniques utilized are new to shock waves and compressible flow in general.

The principle contribution is a new theory for the tangential and normal derivatives of flow properties, such as the pressure and velocity components, just downstream of an infinitesimally thin shock wave. A steady, two-dimensional or axisymmetric shock with a uniform freestream is initially treated. The text concludes with the general case where the upstream flow is nonuniform and the unsteady shock is three-dimensional. In this case, formulation of the shock-based Euler equations required the introduction of the mathematical techniques mentioned above. Consistently, initial data, such as the shape of the shock and upstream flow conditions, is prescribed in a Cartesian coordinate system, as would be the case for computational fluid dynamics (CFD) or experimental data. These data are systemically transformed to provide shock-based derivative results. Numerous applications are discussed. These include derivatives along characteristics, wave reflection from the downstream surface of a shock, intrinsic coordinate derivatives useful for curved shock theory, and so forth. A number of illustrative examples are also provided. The final one is for a curved, unsteady shock that is based on a single Mach reflection shock experiment.

Aside from derivatives, there is an extensive treatment of shock-generated vorticity. A novel analysis of triple points is also included. Both analyses are accompanied with a wide-ranging parametric study.

Acknowledgments

A deep debt of gratitude is owed to Sannu Molder for his insight and stimulation. It is a pleasure to acknowledge the expert typing of Gloria Madden, whose patience with the innumerable revisions was a source of encouragement. The book would not have been possible without the love and understanding of Lita, my wife and soul mate.

1 Introduction

The most distinctive feature of a supersonic flow is shock waves. They were discovered theoretically by Rankin, in 1870, and Hugoniot, in 1877. Ernst Mach was the first to demonstrate their existence by publishing in the 1880s schlieren photographs of a bullet in supersonic flight. Van Dyke (1982) provides an enlargement of one of these photographs. This remarkable picture shows a detached bow shock, a shoulder-based expansion wave, a recompression shock, and a turbulent wake. Nevertheless, shock wave theory developed slowly until World War II. At the time of the war, only the basic fundamentals were known; this material is usually covered in an undergraduate compressible flow course. After the war, the pace of discovery quickened, spurred on by interest in supersonic flight, nuclear explosions, and the reentry physics of long-range missiles.

A range of shock wave topics have been under investigation. These range from the internal structure of a shock to shock wave reflection, refraction, diffraction, and interference (see Emanuel 1986, Chapter 19). Shock waves occur in both steady and unsteady flows. They may be associated with a wide variety of physical phenomena (e.g., chemical reactions that change the shock into a detonation wave). Interaction phenomena are also important, especially shock wave boundary-layer interaction.

A selective treatment of analytical shock wave topics is provided. Computational and experimental topics are not considered, as the author has no expertise in these areas. More comprehensive treatments can be found (e.g., in Ben-Dor 2007; Ben-Dor et al. 2001; Glass and Sislian 1994). Much of the monograph is devoted to establishing and applying the tangential and normal derivatives of various flow properties, just downstream of a shock. The "just downstream" proviso holds throughout the manuscript. In this regard, a cohesive and systematic presentation is provided of these derivatives for a curved shock.

Aside from derivatives, shock-generated vorticity and triple points are prominently treated. Analytical applications, for example, include the determination if a reflected wave from the downstream side of a shock is expansive or compressive. Many other analytically oriented topics are discussed. The primary function of this monograph, hopefully, is to provide an analytical understanding of shock waves.

The first publication to systematically deal with the derivatives of flow variables on the downstream side of a curved, two-dimensional shock is by Thomas (1947). This work was extended by Kanwal (1958b) to a three-dimensional shock. Neither author considers normal derivatives, and both utilize a succinct tensor presentation. In their treatment of the inverse problem for a blunt body flow, Hayes and Probstein (1959) solve for the relevant normal derivatives just downstream of a spherical-shaped (axisymmetric) shock.

The dominant part of this monograph presents the author's analysis for the tangential and normal derivatives just downstream of a curved shock. Much of the analysis is for a two-dimensional or axisymmetric shock. But Chapter 7, a triple-point analysis, and Chapter 9, a derivative analysis, allow for a nonuniform upstream flow and for a three-dimensional shock. Throughout the analysis, the strength of the shock is arbitrary (e.g., the treatment holds for very weak shock waves).

Analytical treatments of the jump conditions, tangential and normal derivatives of common flow properties, such as the pressure, density, and velocity components, are useful in a variety of ways. The resulting equations, including for the vorticity, can be used to check computational fluid dynamics (CFD) solutions (an early intention), and assist in the further development of curved shock theory that has been recently accomplished (see Molder 2012). As pointed out by Hornung (2010), it can be used to treat detonation-type flows. It may be of some use in the further development of shock-capturing CFD schemes.

The discussion presumes familiarity with the basic concepts of a shock wave, including the equations for an oblique shock in a steady flow of a perfect gas. Familiarity with, for example, the theory of characteristics, is also desirable. This text is an extension of Chapter 6 in Emanuel (2001), which is the basis for Chapters 2 through 4. A number of chapters and appendices in this reference also provide background material relevant to our discussion. This is evident by the occasional reference to this book.

The standard jump conditions across an oblique shock are algebraic equations that are independent of time and any coordinate system. At a given point on the shock, they thus hold in an unsteady, three-dimensional flow, where the upstream flow may be nonuniform. However, this generality is limited; it only holds at an instant of time (i.e., a snapshot) when adequate shock data are available. Because a flow field solution is not considered, independent shock data would similarly be required at a different time. If the jump conditions utilize a particular gas model, such as a calorically imperfect gas, the generalized jump equations still apply. "At a given point on the shock" means the analysis is a local one. Central to the analysis is the use of a flow plane (Kaneshige and Hornung 1999), which is defined by the upstream velocity and by a normal vector to the shock's surface, both at the point of interest on a shock. The change in the fluid's momentum, across the shock, is confined to the flow plane.

The jump conditions and tangential derivatives thereby hold under the general conditions mentioned above. This is not the case, however, for the normal derivatives. These cannot be obtained by simply differentiating the jump conditions in a direction normal to the shock, because these relations only hold on the shock's surface. Instead, the normal derivatives require the use of the Euler equations in a shock-based coordinate system. Normal derivative results are therefore more restrictive.

The tangential derivatives are readily obtained when the shock is steady and the upstream flow is uniform. If the upstream flow is nonuniform, these derivatives are altered. If a curved shock is moving into a uniform freestream, the tangential derivatives must account for the upstream variation, along the shock of the shock's velocity. (This point is illustrated in Section 9.10.)

If the freestream or upstream flow is uniform and the overall flow is steady, the flow is automatically homenergetic (constant stagnation enthalpy). A nonuniform freestream, in a steady flow, may be homenergetic or not. (A nonuniform, homenergetic flow example is discussed in Section 8.7.) The assumption of a homenergetic flow is not always a severe restriction. This is because of the substitution principle (Emanuel 2001, Chapter 8). This principle holds for the steady, three-dimensional Euler equations for a perfect gas. The principle keeps invariant the geometry, including that of the streamlines and of any shocks. Also invariant are the pressure and Mach number. A flow field, however, can go from being homenergetic to isoenergetic, where the stagnation enthalpy, which is constant along a streamline, can now vary from streamline to streamline. For instance, a conventional uniform freestream can become a rotational, isoenergetic, parallel flow. The homenergetic assumption is often assumed or is analytically convenient, but can sometimes be removed by invoking the substitution principle.

The next chapter derives general shock jump conditions that hold for an unsteady, three-dimensional flow without assuming a perfect gas or a uniform upstream flow. As an illustration, this chapter concludes with an analysis of an unsteady, normal shock. The third chapter derives shock-based basis vectors, scale factors, and orthogonal, curvilinear coordinates for a generic two-dimensional or axisymmetric shock. Chapter 4 applies the foregoing material to obtain the jump conditions, tangential derivatives along the shock in the flow plane, and the normal derivatives in the downstream direction. Results, with a uniform notation used throughout the rest of the text, are summarized in an appendix.

Much of the analysis in Chapters 2 through 4 comes from Chapter 6 of Emanuel (2001). In turn, this material stems from Emanuel and Liu (1988). Subsequent material, however, is new.

A wide variety of analytical applications of the material in Chapter 4 are discussed in the fifth chapter: for instance, conditions when the shock is normal to the freestream, the Crocco and Thomas points, derivatives along Mach lines and along intrinsic coordinates, flows with convex or concave shocks, and so forth. Typically, results are exact, explicit, algebraic, and readily computer programmable. Shock-generated vorticity and its substantial derivative are the topic of Chapter 6. The chapter concludes with a parametric study of the vorticity and its substantial derivative for a variety of freestream Mach numbers for generic, two-dimensional, or axisymmetric shocks. A side benefit illustrates the three-dimensional relief effect associated with an axisymmetric flow. A second break is made in Chapter 7 from the emphasis on derivatives. Instead, a novel treatment of a triple-point is provided. These occur with a lambda shock wave system, which is a relatively common flow configuration. Chapter 8 generalizes the Chapter 4 analysis of a two-dimensional or axisymmetric shock when the upstream flow is nonuniform. A simple model illustrates the analysis.

The last chapter provides a general derivative formulation that often uses an operator approach to simplify derivative evaluations. In the second section, a shock-based, orthogonal basis system is introduced. The analysis in this section holds for a three-dimensional, unsteady shock, the gas need not be perfect, and the upstream flow may be nonuniform, including being rotational and nonhomenergetic. The third section introduces a steady, elliptic paraboloid shock, with a uniform freestream (for analytical simplicity) that is used to illustrate the analysis in Sections 9.4 through 9.8, where the shock is steady. In the limit of the elliptic paraboloid shock becoming two-dimensional or axisymmetric, results are obtained that are compared with earlier material, thereby partially validating the more general analysis. (This "bootstrap" approach was essential for the development of the normal derivative analysis in Section 9.7.) The next section evaluates the curvatures of the shock in the flow plane and in a plane normal to the shock and the flow plane. The fifth section evaluates the vorticity on the downstream side of the shock, but the flow is now assumed to be a perfect gas and is also homenergetic. The perfect gas assumption is utilized in the rest of the chapter, but not the homenergetic assumption. Section 9.6 obtains the jump conditions and the tangential derivatives in the two orthogonal planes mentioned above. An approach similar to that in Chapter 3 failed to produce a global, shock-based coordinate system. As in Chapter 4, this system was to be used with the Euler equations to obtain the normal derivatives. Appendix K shows, by means of several examples, that the desired coordinate system does not exist for a three-dimensional shock.

Instead, in Section 9.7, a new technique is provided for formulating the steady Euler equations in a curvilinear, shock-based coordinate system. These equations then provide the general relations for the normal derivatives. This method can be viewed as a local analysis in contrast to unsuccessful global approaches. Results are summarized in three appendices, where the last one is for an elliptic paraboloid shock. Section 9.8 discusses a few analytical applications, such as normal derivatives when the shock is normal to its upstream velocity. Also provided is a discussion of a general intrinsic coordinate basis.

Section 9.9 provides the analysis for the case where the upstream flow is non-uniform and the unsteady shock is three-dimensional. The approach is an extension of that in Section 9.7. The final section applies the unsteady theory to a curved, reflected shock that is part of a single Mach reflection pattern. The data used stem from a shock tube experiment.

Following the chapter, presented in the following order, are the appendices, problems, and references. Some of the problems are an integral part of the analysis and are referred to in the text. There are many appendices—the first is a selective nomenclature for the more frequently encountered symbols. Some of these appendices are of a summary nature designed to make the analysis more reader- and user-friendly.

2 General Jump Conditions

2.1 BASIS VECTOR SYSTEM AND SHOCK VELOCITY

Only the fundamental assumption of a continuum flow with an infinitesimally thin shock is pertinent to this section. A fixed Cartesian coordinate system x_i and its corresponding orthonormal basis $\hat{I}_i$ are introduced. The shock wave surface, which may be in motion, is represented by

$$F = F(x_i, t) = 0 \tag{2.1}$$

Conditions just upstream and just downstream of the surface are denoted with subscripts 1 and 2, respectively. In a more conventional treatment, the upstream flow is uniform and steady, and the "just upstream" qualification is unnecessary. The velocity, in a laboratory frame, just upstream and downstream of the shock, is written as

$$\vec{V}_j = V_{j,i}(x_k, t)\,\hat{I}_i, \quad j = 1, 2 \tag{2.2}$$

where x_k and t satisfy Equation (2.1), and i is summed over. The arbitrary sign of F is chosen so that

$$\vec{V}_1 \cdot \nabla F \geq 0 \tag{2.3}$$

for some region of the shock's surface. For this region, the flow is primarily in the downstream direction.

A unit vector $\hat{n}$ is defined as

$$\hat{n} = \hat{n}(x_i, t) = \frac{\nabla F}{|\nabla F|} \tag{2.4}$$

which is normal to the shock and, in view of Equation (2.3), is oriented in the downstream direction. The shock wave's velocity, $\vec{V}_s$, is introduced that is normal to its surface. As will become evident, a tangential shock wave velocity component is not defined. This normal velocity is obtained by setting the substantial derivative of the surface equal to zero

$$\frac{\partial F}{\partial t} + \vec{V}_s \cdot \nabla F = 0$$

With $\vec{V}_s$ proportional to $\hat{n}$, we obtain

$$\vec{V}_s = V_s \hat{n} = -\frac{1}{|\nabla F|}\frac{\partial F}{\partial t}\,\hat{n} \tag{2.5}$$

From the viewpoint of the shock, only the velocity of the gas $\vec{V}_j^*$ relative to it is significant. These velocities are defined by

$$\vec{V}_j^* = \vec{V}_j - \vec{V}_s = \vec{V}_j - V_s \hat{n}, \quad j = 1, 2 \tag{2.6}$$

When $V_s < 0$, the shock is moving into the upstream flow, and there is an increase in the component of $\vec{V}_j^*$ that is normal to the shock.

The vectors $\hat{n}$ and $\vec{V}_1^*$ define a unique plane, called the *flow plane* (Kaneshige and Hornung 1999). Each point of the shock contains such a plane. One exception is when the shock is normal to $\vec{V}_1^*$, which is discussed later. Momentum considerations show that $\vec{V}_2^*$ lies in this plane. Equation (2.6) then shows that the $\vec{V}_j$ also lie in the flow plane.

A unit vector $\hat{t}$ is defined that is tangent to the shock in the flow plane, as sketched in Figure 2.1. We have a right-handed, orthonormal basis $\hat{t}, \hat{n}, \hat{b}$ where the binormal $\hat{b}$ is perpendicular to $\hat{n}$ and $\hat{t}$. It is given by $\hat{t} \times \hat{n}$ and points into the page. This basis moves with the shock, where $\hat{t}$ and $\hat{n}$ remain in the flow plane. The basis is designed to become the $\hat{l}_i$ basis by a solid-body rotation. Suppose the shock is normal to $\vec{V}_1^*$, then, at this point, a solid-body rotation yields

$$\hat{n} = \hat{l}_1, \quad \hat{t} = \hat{l}_2, \quad \hat{b} = -\hat{l}_3$$

where $\vec{V}_1^* = V_1^* \hat{l}_1$.

Starting in Chapter 3, a right-handed, orthogonal coordinate system, ξ_i, is introduced where ξ_1 is tangent to $\hat{t}$, ξ_2 to $\hat{n}$, and ξ_3 to $\hat{b}$. It is especially convenient for ξ_1 to be tangent to $\hat{t}$ to expedite the normal derivative analysis in Section 4.4. (This is the reason for using $\hat{t}, \hat{n}, \hat{b}$ as the orthogonal basis rather than $\hat{n}, \hat{t}, \hat{b}$ where $\hat{b} = \hat{l}_3$.) Once the h_i scale factors are obtained in Chapter 3, the ξ_1, ξ_2, ξ_3 coordinates are replaced with s, n, b coordinates, respectively.

The flow plane definition enables us to introduce the θ and β angles. These are conventionally used with a steady, planar, oblique shock wave and a uniform upstream flow. Figure 2.1 is a sketch for a convex shock, relative to the freestream, and shows θ as the acute angle between $\vec{V}_1^*$ and $\vec{V}_2^*$ and β as the acute angle between $\vec{V}_1^*$ and the shock. Both angles are in the first quadrant. In Sections 5.1 and 5.5,

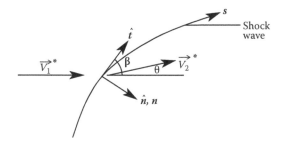

FIGURE 2.1 Section through a shock that contains both $\vec{V}_1^*$ and $\hat{n}$ vectors (i.e., the flow plane). The $\hat{b}$ vector is normal to the plane of the page and points into the page.

a concave shock, relative to the freestream, is discussed. In the concave case, θ is negative and in the fourth quadrant, while β is in the second quadrant. In general, their values may change from point to point, or with time at a given point, on the shock's surface. When the shock is normal to $\vec{V}_1^*$, θ and β are $0°$ and $90°$, respectively.

Although F and $\vec{V}_1^*$ are presumed known in terms of a Cartesian coordinate system, the $\hat{t},\hat{n},\hat{b}$ system is far more convenient for the analysis. For instance, from Figure 2.1, we obtain

$$\vec{V}_1^* = V_1^*(\sin\beta\hat{n} + \cos\beta\hat{t}) \tag{2.7a}$$

$$\vec{V}_2^* = V_2^*[(\sin(\beta-\theta)\hat{n} + \cos(\beta-\theta)\hat{t}] \tag{2.7b}$$

while Equations (2.6) yield

$$\vec{V}_1 = (V_1^*\sin\beta + V_s)\hat{n} + V_1^*\cos\beta\hat{t} \tag{2.8a}$$

$$\vec{V}_2 = [V_2^*\sin(\beta-\theta) + V_s]\hat{n} + V_2^*\cos(\beta-\theta)\hat{t} \tag{2.8b}$$

Alternatively, an explicit form for $\vec{V}_1^*$ is

$$\vec{V}_1^* = \vec{V}_1 - V_s\hat{n} = V_{1,i}\hat{i}_i + \frac{\partial F}{\partial t}\frac{\nabla F}{|\nabla F|^2} \tag{2.9}$$

where the quantities on the rightmost side are known functions of x_i and t. We therefore view $\vec{V}_1^*$ as a known velocity.

The binormal basis vector is obtained by taking the cross product of $\vec{V}_1^*$ with $\hat{n}$:

$$\hat{b} = -\frac{\hat{n} \times \vec{V}_1^*}{V_1^*\cos\beta} \tag{2.10a}$$

where the denominator converts $\hat{n} \times \vec{V}_1^*$ into a unit vector. The $\hat{t}$ vector is given by

$$\hat{t} = -\hat{b} \times \hat{n} = \frac{1}{\cos\beta}\left(\frac{\vec{V}_1^*}{V_1^*} - \sin\beta\,\hat{n}\right) \tag{2.10b}$$

which reduces to an identity with the use of Equation (2.7a). With the aid of Equations (2.4) and (2.7a), β is given by

$$\sin\beta = \frac{\vec{V}_1^* \cdot \nabla F}{V_1^*|\nabla F|} \tag{2.11}$$

A relation for θ is discussed in Section 2.3, "Explicit Solution."

Parameters $\hat{b}$, $\hat{t}$, and $\hat{n}$ have been evaluated in terms of $\vec{V}_1^*$. It is also useful to obtain these parameters in terms of $\vec{V}_1$. We write $\hat{b}$ as

$$\hat{b} = -\frac{\hat{n} \times \vec{V}_1^*}{V_1^* \cos\beta} = -\frac{\hat{n} \times (\vec{V}_1 - V_s \hat{n})}{V_1^* \cos\beta} = -\frac{\hat{n} \times \vec{V}_1}{V_1^* \cos\beta} = -\frac{1}{I} \nabla F \times \vec{V}_1 \qquad (2.12)$$

where

$$I = V_1^* \mid \nabla F \mid \cos\beta = [V_1^{*2} \mid \nabla F \mid^2 - (\vec{V}_1^* \cdot \nabla F)^2]^{1/2}$$

in view of Equation (2.11). With the derivative notation

$$F_t = \frac{\partial F}{\partial t}, \quad F_{x_i} = \frac{\partial F}{\partial x_i}$$

and Equation (2.9), we obtain

$$V_{1,i}^* = V_{1,i} + \frac{F_t F_{x_i}}{\mid \nabla F \mid^2}$$

$$\vec{V}_1^* \cdot \nabla F = \vec{V}_1 \cdot \nabla F + F_t$$

Hence, I becomes

$$I = \left[\left(V_1^2 + 2\frac{F_t}{\mid \nabla F \mid^2} \vec{V}_1 \cdot \nabla F + \frac{F_t^2}{\mid \nabla F \mid^2} \right) \mid \nabla F \mid^2 - \left(\vec{V}_1 \cdot \nabla F \right)^2 - 2F_t \vec{V}_1 \cdot \nabla F - F_t^2 \right]^{1/2}$$

$$= \left[V_1^2 \mid \nabla F \mid^2 - \left(\vec{V}_1 \cdot \nabla F \right)^2 \right]^{1/2}$$

$$(2.13)$$

By comparison with the first I equation, we observe that V_1^*, $\vec{V}_1^*$ can be replaced with V_1, $\vec{V}_1$. Finally, note that $\hat{t}$ can be written as

$$\hat{t} = -\hat{b} \times \hat{n} = -\frac{1}{I \mid \nabla F \mid} (\nabla F \times \vec{V}_1) \times \nabla F = -\frac{1}{I \mid \nabla F \mid} \left[\mid \nabla F \mid^2 \vec{V}_1 - (\vec{V}_1 \cdot \nabla F) \nabla F \right] \qquad (2.14)$$

Consequently, the $\hat{t}, \hat{n}, \hat{b}$ basis and β can be defined using either $\vec{V}_1^*$ or $\vec{V}_1$.

2.2 CONSERVATION EQUATIONS

The same principles that yield the governing Euler equations are applied to a differential volume element that contains a piece of the shock. Application of these principles then results in the jump conditions, whereby flow conditions on the two sides of the shock are initially symmetrically related. For these equations, the substantial derivative is required:

$$\left(\frac{DF}{Dt}\right)_j = \frac{\partial F}{\partial t} + \vec{V}_j \cdot \nabla F, \quad j = 1,2 \tag{2.15}$$

where the velocity is for a fluid particle in a laboratory frame.

Conservation of the flux of mass across the shock, in a laboratory frame, is given by

$$\left(\rho \frac{DF}{Dt}\right)_1 = \left(\rho \frac{DF}{Dt}\right)_2 \tag{2.16}$$

where ρ is the density. This relation can be understood by writing the upstream side as

$$\left(\rho \frac{DF}{Dt}\right)_1 = \rho_1 \left(\frac{\partial F}{\partial t} + \vec{V}_1 \cdot \nabla F\right) = \rho_1 |\nabla F| \left(\frac{1}{|\nabla F|}\frac{\partial F}{\partial t} + \hat{n} \cdot \vec{V}_1\right)$$

with a similar result for the downstream side. Mass flux conservation now becomes

$$\frac{\rho_1}{|\nabla F|}\frac{\partial F}{\partial t} + \rho_1 \hat{n} \cdot \vec{V}_1 = \frac{\rho_2}{|\nabla F|}\frac{\partial F}{\partial t} + \rho_2 \hat{n} \cdot \vec{V}_2$$

The two $\hat{n}$ terms represent the mass flux across the shock, as if it were steady, while the two $\partial F/\partial t$ terms provide the contribution from a moving shock. Recall that $|\nabla F|^{-1}(\partial F/\partial t)$ also appeared in Equation (2.5), where it represents the normal component of the velocity of a moving shock.

In a similar manner, equations are written that represent, across the shock, the normal component of momentum, the tangential momentum component, and the energy. We thereby obtain

$$\left[p|\nabla F|^2 + \rho\left(\frac{DF}{Dt}\right)^2\right]_1 = \left[p|\nabla F|^2 + \rho\left(\frac{DF}{Dt}\right)^2\right]_2 \tag{2.17}$$

$$\left(\rho\vec{V}\cdot\hat{t}\frac{DF}{Dt}\right)_1 = \left(\rho\vec{V}\cdot\hat{t}\frac{DF}{Dt}\right)_2 \tag{2.18}$$

$$\left[h|\nabla F|^2 + \frac{1}{2}\left(\frac{DF}{Dt}\right)^2\right]_1 = \left[h|\nabla F|^2 + \frac{1}{2}\left(\frac{DF}{Dt}\right)^2\right]_2 \tag{2.19}$$

where p and h are the pressure and enthalpy. Equations (2.16) through (2.19) are the symmetrical jump conditions in a general form. (In addition, the second law requires the entropy condition, $S_2 \geq S_1$. We do not list it, because it is not directly utilized in the subsequent analysis.) This form is not a convenient one. Explicit equations for the unknowns p_2, ρ_2, h_2, and V_2^* are desired.

With the aid of Equations (2.4) through (2.6), the substantial derivatives that appear in the jump conditions now become

$$\left(\frac{DF}{Dt}\right)_j = \frac{\partial F}{\partial t} + \left(\vec{V}_j^* + V_s\hat{n}\right) \cdot \nabla F = \frac{\partial F}{\partial t} + \vec{V}_j^* \cdot \nabla F - \frac{1}{|\nabla F|}\frac{\partial F}{\partial t}\frac{\nabla F \cdot \nabla F}{|\nabla F|} = \vec{V}_j^* \cdot \nabla F, \quad j=1,2$$

Hence, the conditions simplify to

$$\left(\rho\vec{V}^* \cdot \hat{n}\right)_1 = \left(\rho\vec{V}^* \cdot \hat{n}\right)_2 \tag{2.20a}$$

$$\left[p + \rho\left(\vec{V}^* \cdot \hat{n}\right)^2\right]_1 = \left[p + \rho\left(\vec{V}^* \cdot \hat{n}\right)^2\right]_2 \tag{2.20b}$$

$$\left(\vec{V}^* \cdot \hat{t}\right)_1 = \left(\vec{V}^* \cdot \hat{t}\right)_2 \tag{2.20c}$$

$$\left[h + \frac{1}{2}\left(\vec{V}^* \cdot \hat{n}\right)^2\right]_1 = \left[h + \frac{1}{2}\left(\vec{V}^* \cdot \hat{n}\right)^2\right]_2 \tag{2.20d}$$

These are still symmetrical jump conditions, but now in a frame fixed to the shock. They have a more familiar appearance as compared to the preceding laboratory frame version. We could have started with these relations in preference to Equations (2.16) through (2.19).

2.3   EXPLICIT SOLUTION

In order to evaluate the dot products that appear in the equations in Equation (2.20), Equation (2.7) is utilized, with the result

$$\vec{V}_1^* \cdot \hat{t} = V_1^* \cos\beta \tag{2.21a}$$

$$\vec{V}_1^* \cdot \hat{n} = V_1^* \sin\beta \tag{2.21b}$$

$$\vec{V}_2^* \cdot \hat{t} = V_2^* \cos(\beta - \theta) \tag{2.21c}$$

$$\vec{V}_2^* \cdot \hat{n} = V_2^* \sin(\beta - \theta) \tag{2.21d}$$

Equation (2.20c) for the velocity tangency condition yields

$$V_2^* = V_1^* \frac{\cos\beta}{\cos(\beta - \theta)} \tag{2.22}$$

This relation cannot be used for a normal shock, because the ratio of cosines is indeterminant. (The normal shock formulation is discussed in the subsequent illustrative example.)

Conservation of mass flux, Equation (2.20a) yields

$$\rho_1 V_1^* \sin\beta = \rho_2 V_2^* \sin(\beta - \theta)$$

In combination with Equation (2.22), this becomes

$$\rho_2 = \rho_1 \frac{\tan\beta}{\tan(\beta - \theta)} \tag{2.23}$$

Equation (2.20b) produces

$$p_2 = p_1 + \left(\rho V^{*2}\right)_1 \sin^2\beta - \left(\rho V^{*2}\right)_2 \sin^2(\beta - \theta)$$

$$= p_1 + \left(\rho V^{*2}\right)_1 \left[\sin^2\beta - \frac{\sin\beta\cos\beta\sin(\beta - \theta)}{\cos(\beta - \theta)}\right]$$

$$= p_1 + \left(\rho V^{*2}\right)_1 \frac{\sin\beta\sin\beta}{\cos(\beta - \theta)} \tag{2.24}$$

where Equations (2.22) and (2.23) are used. The energy equation now becomes

$$h_2 = h_1 + \frac{1}{2} V_1^{*2} \sin^2\beta - \frac{1}{2} V_2^{*2} \sin^2(\beta - \theta)$$

$$= h_1 + \frac{1}{2} V_1^{*2} \left[\sin^2\beta - \frac{\cos^2\beta\sin^2(\beta - \theta)}{\cos^2(\beta - \theta)}\right]$$

$$= h_1 + \frac{1}{2} V_1^{*2} \frac{\sin(2\beta - \theta)\sin\theta}{\cos^2(\beta - \theta)} \tag{2.25}$$

Downstream variables V_2^*, ρ_2, p_2, and h_2 are explicitly provided by these equations. The equations hold for unsteady, three-dimensional shocks and do not assume a perfect gas; note the absence of the ratio of specific heats. Variables are not normalized, because their upstream counterparts are functions of position on the shock surface and time. The downstream velocity $\vec{V}_2$ is then provided by Equation (2.8b). The parameters on the right sides consist of β, θ, ρ_1, p_1, h_1, and V_1^*. Except for θ, these quantities are presumed known, where β is given by Equation (2.11).

To evaluate θ, a thermodynamic state equation involving ρ, p, and h needs to be introduced. Because the enthalpy is present in only one jump condition, the most convenient form for this relation is

$$h = h(p,\rho) \tag{2.26}$$

Equation (2.25) thus becomes

$$h(p_2,\rho_2) = h(p_1,\rho_1) + \frac{1}{2}V_1^{*2}\frac{\sin\theta\sin(2\beta-\theta)}{\cos^2(\beta-\theta)} \tag{2.27}$$

Equations (2.23) and (2.24) can now be used to eliminate ρ_2 and p_2 from $h(p_2,\rho_2)$. The result would be an implicit equation for θ. Alternate approaches for treating real gas shock wave phenomena are provided by Vincenti and Kruger (1965), Zel'dovich and Raizer (1966), and Zucrow and Hoffman (1976).

For a perfect gas, Problem 1 shows that Equation (2.27) reduces to the conventional oblique shock equation:

$$\tan\theta = \cot\beta\frac{M_1^2\sin^2\beta-1}{1+\left\{\left[(\gamma+1)/2\right]-\sin^2\beta\right\}M_1^2} \tag{2.28}$$

where the upstream Mach number is

$$M_1 = \frac{V_1^*}{(\gamma p_1/\rho_1)^{1/2}}$$

In this case, the equation for θ is explicit. An explicit result for β is provided by Appendix B, where Equation (2.28) is inverted. Problem 2 develops the van der Waals state equation counterpart to Equation (2.28).

The above analysis holds for both convex and concave shocks. Note that the ratio factor in Equation (2.28) is nonnegative. The previous angle convention, for both types of shocks, is such that the product, $\tan\theta\tan\beta$, is nonnegative, regardless of the shock's orientation.

2.4   ILLUSTRATIVE EXAMPLE

A shock wave may be unsteady for any of several reasons, such as an unsteady upstream flow. A weak source of unsteadiness would stem from a turbulent boundary layer, as sketched in Figure 2.2. Disturbances generated by the boundary layer travel along Mach waves until these waves impinge on the shock. As indicated in the figure, the Mach waves from both the upstream and downstream walls travel toward the shock.

If a shock is sufficiently intense, the flow downstream of it is subsonic. In this circumstance, disturbances can propagate in an upstream direction, thereby causing the shock to become unsteady. This mechanism is involved with the buzz phenomenon

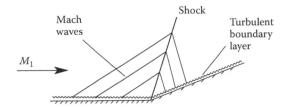

FIGURE 2.2 Oblique shock caused by a sharp wall turn. The Mach waves emanate from a turbulent layer.

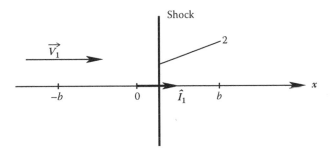

FIGURE 2.3 An unsteady normal shock.

of a jet engine inlet in supersonic flight. For instance, consider an axisymmetric, supersonic inlet with a single centrally located cone. During buzz, which typically occurs with a frequency of about 10 to 20 Hz (Sterbenz and Evvard, 1955), there is a single, detached, nearly normal shock when the shock is in its most upstream position. When in its preferred downstream position, it is a multiple system of oblique shock waves, where the upstream-most shock is conical and is attached to the apex of the cone.

To illustrate the theory, a sinusoidal oscillation

$$F = x - b\sin(2\pi \kappa t) = 0$$

of a normal shock is examined, where the amplitude b and frequency κ are constants. The upstream supersonic velocity (see Figure 2.3)

$$\vec{V}_1 = V_1 \hat{l}_1 \tag{2.29}$$

is taken as steady and uniform.

For this flow, note that $\hat{n} = \hat{l}_1$,

$$\vec{V}_1 \cdot \nabla F = V_1 > 0$$

in accordance with Equation (2.3), and $\hat{t}$ and $\hat{b}$ are unnecessary. The shock velocity is

$$\vec{V}_s = V_s \hat{l}_1 = -\frac{\hat{n}}{|\nabla F|}\frac{\partial F}{\partial t} = 2\pi b \kappa \cos(2\pi \kappa t)\hat{l}_1$$

When the shock is moving to the right, $V_s > 0$ and M_2 exceeds the value

$$\bar{M}_2 = \left[\frac{1+(\gamma-1)M_1^2/2}{\gamma M_1^2-(\gamma-1)/2}\right]^{1/2}, \quad V_s=0 \tag{2.30}$$

it would have if the shock were steady. (See Equations 2.34 for Mach number definitions.) Thus, when $V_s > 0$, the shock is weaker than if it were stationary. However, $M_2 - M_s$ cannot exceed unity, because the shock's motion is caused by a disturbance in the downstream flow, which is subsonic relative to the shock.

When the shock is moving to the left, $V_s < 0$, it is stronger than its stationary value, and $M_2 < \bar{M}_2$. Furthermore, M_2 may be negative if V_s is sufficiently negative. The M_1, $\bar{M}_2$, and M_2 Mach numbers are with respect to a laboratory frame, not a shock-fixed frame.

For a normal shock $\theta = 0°$, $\beta = 90°$, and Equations (2.22) through (2.25) are indeterminate. This difficulty is avoided by using the initial equations for ρ_2, p_2, and h_2. For example, ρ_2 is given by

$$\rho_2 = \rho_1 \frac{V_1 - V_s}{V_2 - V_s} \tag{2.31a}$$

Similarly, p_2 and h_2 are given by

$$p_2 = p_1 + \rho_1(V_1 - V_s)(V_1 - V_2) \tag{2.31b}$$

$$h_2 = h_1 + \frac{1}{2}V_1^{*2} - \frac{1}{2}V_2^{*2} = h_1 + \frac{1}{2}\left(V_1^* - V_2^*\right)\left(V_1^* + V_2^*\right)$$

$$= h_1 + \frac{1}{2}\left(V_1 - V_2\right)\left(V_1 + V_2 - 2V_s\right) \tag{2.31c}$$

To proceed with the analysis, a perfect gas is now assumed and the enthalpy equation becomes

$$\frac{p_2}{\rho_2} = \frac{p_1}{\rho_1} + \frac{\gamma-1}{2\gamma}\left(V_1 - V_2\right)\left(V_1 + V_2 - 2V_s\right) \tag{2.32a}$$

From Equations (2.31a) and (2.31b), we have

$$\frac{p_2}{\rho_2} = \frac{p_1}{\rho_1}\frac{V_2 - V_s}{V_1 - V_s} + \left(V_1 - V_2\right)\left(V_2 - V_s\right) \tag{2.32b}$$

After eliminating p_2/ρ_2, an explicit solution for V_2 is obtained:

$$V_2 = \frac{\gamma-1}{\gamma+1}V_1 + \frac{2\gamma}{\gamma+1}\frac{p_1/\rho_1}{V_1 - V_s} + \frac{2}{\gamma+1}V_s \tag{2.33}$$

Thus, V_2 equals a constant term, a V_s term that is proportional to $\cos(2\pi\kappa t)$, and a term with this cosine in a denominator, which dominates when the denominator is small.

As usual with a perfect gas, it is convenient to introduce the Mach numbers:

$$M_1 = \frac{V_1}{a_1} = \frac{V_1}{\left(\gamma p_1/\rho_1\right)^{1/2}} \tag{2.34a}$$

$$M_2 = \frac{V_2}{a_2} = \frac{V_2}{\left(\gamma p_2/\rho_2\right)^{1/2}} \tag{2.34b}$$

$$M_s = \frac{V_s}{a_1} = 2\pi\left(\frac{\rho_1}{\gamma p_1}\right)^{1/2} b\kappa\cos(2\pi\kappa t) \tag{2.34c}$$

where M_s, the shock wave Mach number, is negative whenever V_s is negative. In order to obtain explicit results, a relation is needed between the upstream and downstream sound speeds. Multiply Equation (2.32a) with γ to obtain

$$a_2^2 = a_1^2 + \frac{\gamma-1}{2}(V_1 - V_2)(V_1 + V_2 - 2V_s)$$

With the aid of Equation (2.33), V_2 is eliminated with the result

$$\frac{a_2^2}{a_1^2} = 1 + \frac{2(\gamma-1)}{(\gamma+1)^2}\frac{\left[(M_1-M_s)^2-1\right]\left[\gamma(M_1-M_s)^2+1\right]}{(M_1-M_s)^2}$$

or

$$\frac{a_2}{a_1} = \frac{2}{(\gamma+1)}\frac{\left[\gamma(M_1-M_s)^2-\frac{\gamma-1}{2}\right]^{1/2}\left[1+\frac{\gamma-1}{2}(M_1-M_s)^2\right]^{1/2}}{(M_1-M_s)} \tag{2.35}$$

This is the usual jump condition formula for the speed of sound ratio with M_1 replaced by $M_1 - M_s$.

Equation (2.33) is now written as

$$a_2 M_2 = \frac{\gamma-1}{\gamma+1}a_1 M_1 + \frac{2}{\gamma+1}\frac{a_1}{M_1-M_s} + \frac{2}{\gamma+1}a_1 M_s$$

or

$$M_2 = \frac{2}{\gamma+1}\frac{1+(M_1-M_s)\left(\frac{\gamma-1}{2}M_1+M_s\right)}{M_1-M_s}\frac{a_1}{a_2}$$

Equation (2.35) is utilized to eliminate a_1/a_2, with the result

$$M_2 = \frac{1+(M_1 - M_s)\left(\frac{\gamma-1}{2}M_1 + M_s\right)}{\left[1+\frac{\gamma-1}{2}(M_1 - M_s)^2\right]^{1/2}\left[\gamma(M_1 - M_s)^2 - \frac{\gamma-1}{2}\right]^{1/2}} \qquad (2.36)$$

which reduces to Equation (2.30) when $M_s = 0$. This relation provides the time dependence of M_2 through M_s, which is given by Equation (2.34c). While the shock speed and M_s are simple sinusoids, the variation of M_2, w_2, p_2, ..., are not as simple.

From the denominator of Equation (2.36), a real solution for M_2 requires

$$(M_1 - M_s)^2 > \frac{\gamma-1}{2\gamma}$$

Since $M_1 - M_s$ can be written as

$$M_1 - M_s = M_1\left[1 - 2\pi\left(\frac{b\kappa}{V_1}\right)\cos(2\pi\kappa t)\right]$$

the left side of the inequality is a minimum when the cosine is unity. As a consequence, the inequality can be written as

$$\left(1 - 2\pi\frac{b\kappa}{V_1}\right)^2 > \frac{\gamma-1}{2\gamma}\frac{1}{M_1^2}$$

The right side is always well below unity, and small values for $b\kappa/V_1$ readily satisfy the inequality. Nevertheless, there is a range of values for $2\pi(b\kappa/V_1)$, centered about unity, for which a real solution is not obtained, and the postulated sinusoidal shock motion cannot occur. The $V_1 - V_s$ denominator in Equation (2.33), which is proportional to $M_1 - M_s$, is therefore limited in how small it can become.

3 Two-Dimensional or Axisymmetric Formulation

3.1 BASIS VECTORS

We assume a steady, two-dimensional ($\sigma = 0$) or axisymmetric ($\sigma = 1$) flow of a perfect gas that contains a shock wave. In addition, no sweep or swirl and a uniform upstream flow are assumed. A Cartesian coordinate system initially is utilized, as sketched in Figure 3.1, where x_1 is aligned with the uniform freestream velocity. It is convenient to introduce a transverse radial position vector:

$$\vec{R} = x_2 \hat{\imath}_2 + \sigma x_3 \hat{\imath}_3 \tag{3.1a}$$

where

$$R = \left(x_2^2 + \sigma x_3^2 \right)^{1/2}, \quad \frac{\partial R}{\partial x_2} = \frac{x_2}{R}, \quad \frac{\partial R}{\partial x_3} = \frac{\sigma x_3}{R} \tag{3.1b}$$

and its normalized form is

$$\hat{\varepsilon}_R = \frac{\vec{R}}{R} = \frac{x_2}{R} \hat{\imath}_2 + \frac{\sigma x_3}{R} \hat{\imath}_3 \tag{3.1c}$$

The constant freestream velocity is given by Equation (2.29).

For the derivative analysis, a known shock shape is presumed. Of course, from a computational fluid dynamics (CFD) point of view, the shock's location is generally not known but must be found. For our purposes, however, the assumption is warranted, and the resulting jump and derivative relations hold, whether or not the shock's location is actually known.

The shape of the two-dimensional or axisymmetric shock is written as

$$F = f(x_1) - R = 0 \tag{3.2}$$

The gradient of F and its magnitude are

$$\nabla F = \frac{df}{dx_1} \hat{\imath}_1 - \frac{x_2}{R} \hat{\imath}_2 - \frac{\sigma x_3}{R} \hat{\imath}_3 = f' \hat{\imath}_1 - \hat{\varepsilon}_R \tag{3.3a}$$

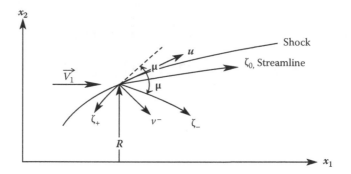

FIGURE 3.1 Streamline (ζ_0), left-running (ζ_+), and right-running (ζ_-) characteristic directions on the downstream side of a shock.

$$|\nabla F| = \left(f'^2 + \frac{x_2^2}{R^2} + \frac{\sigma x_3^2}{R^2} \right)^{1/2} = \left(1 + f'^2 \right)^{1/2} \tag{3.3b}$$

where $f' = (df/dx_1)$. Equation (3.2) is an idealized shock shape in that other disturbances, such as a downstream wave that overtakes part of the shock, are not considered. If such an interaction occurs, then Equation (3.2) still holds for the undisturbed part of the shock.

As noted earlier, the normal vector is

$$\hat{n} = \frac{\nabla F}{|\nabla F|} = \frac{f' \hat{l}_1 - \hat{\varepsilon}_R}{\left(1 + f'^2 \right)^{1/2}} \tag{3.4}$$

Its dot product with $\hat{l}_1$ yields the β angle:

$$\hat{l}_1 \cdot \hat{n} = \sin\beta = \frac{f'}{\left(1 + f'^2 \right)^{1/2}} \tag{3.5a}$$

We thus have

$$\cos\beta = \frac{1}{\left(1 + f'^2 \right)^{1/2}}, \quad \tan\beta = f' \tag{3.5b,c}$$

where the $\tan\beta$ result is evident from Equation (3.2). Write $\vec{b}$ as

$$\vec{b} = -\hat{n} \times \vec{V}_1 = -\frac{V_1}{\left(1 + f'^2 \right)^{1/2}} \hat{l}_1 \times \hat{\varepsilon}_R$$

where

$$\hat{1}_1 \times \hat{\varepsilon}_R = \frac{x_2}{R} \hat{1}_1 \times \hat{1}_2 + \frac{\sigma x_3}{R} \hat{1}_1 \times \hat{1}_3 = -\frac{1}{R}\left(\sigma x_3 \hat{1}_2 - x_2 \hat{1}_3\right)$$

This yields

$$\vec{b} = \frac{V_1}{R\left(1 + f'^2\right)^{1/2}}\left(\sigma x_3 \hat{1}_2 - x_2 \hat{1}_3\right)$$

and its normalized value

$$\hat{b} = \frac{1}{R}\left(\sigma x_3 \hat{1}_2 - x_2 \hat{1}_3\right) \tag{3.6}$$

Equation (2.10b) then yields

$$\hat{t} = \frac{1}{\left(1 + f'^2\right)^{1/2}}\left(\hat{1}_1 + f'\hat{\varepsilon}_R\right) \tag{3.7}$$

In order to evaluate quantities such as $\nabla \cdot \vec{V}$ or $\nabla \times \vec{V}$, a curvilinear, orthogonal coordinate system ξ_i is introduced, where ξ_1 is tangent to $\hat{t}$, ξ_3 is tangent to $\hat{b}$, and ξ_2 is tangent to $\hat{n}$. Thus, ξ_1 and ξ_3 are shock surface coordinates, while ξ_1 and ξ_2 are in the flow plane. The velocity and gradient operator are

$$\vec{V} = u\hat{t} + v\hat{n} \tag{3.8}$$

$$\nabla = \frac{\hat{e}_1}{h_1}\frac{\partial}{\partial \xi_1} + \frac{\hat{e}_2}{h_2}\frac{\partial}{\partial \xi_2} + \frac{\hat{e}_3}{h_3}\frac{\partial}{\partial \xi_3} \tag{3.9}$$

where the h_i are scale factors, and

$$\hat{e}_1 = \hat{t}, \quad \hat{e}_2 = \hat{n}, \quad \hat{e}_3 = \hat{b} \tag{3.10}$$

Although $\vec{V}$ has no $\hat{b}$ component and any scalar has

$$\frac{\partial \varphi}{\partial \xi_3} = 0$$

it is necessary to retain the $\hat{b}$ term in the del operator, because $\partial \hat{b}/\partial \xi_3$ is not zero in an axisymmetric flow.

3.2   SHOCK-BASED CURVILINEAR COORDINATES

The task of developing the transformation

$$\xi_j = \xi_j(x_i) \tag{3.11}$$

is not trivial. For ξ_2, however, a simple choice is

$$\xi_2 = F = f(x_1) - R \tag{3.12}$$

where ξ_2 is zero on the shock's surface and, by Equation (2.3), is positive downstream of it.

In a shock produced vorticity analysis by Hayes (1957), a unit vector is introduced that is the gradient of the normal coordinate

$$\mathbf{n} = \nabla \xi_2$$

which results in the identity

$$\nabla \times \mathbf{n} = 0$$

In our formulation, this approach cannot be utilized. It is inconsistent with Equations (2.4) and (3.12) unless $|\nabla F| = 1$. This is not possible, in view of Equation (3.3b), because $f' \neq 0$. Both approaches are correct and can be shown to yield the same equation for $\hat{n}$. The approach used here, however, results in a simpler equation for ξ_2. Hayes' approach requires normalizing ξ_2.

For an arbitrary point, the two coordinate systems yield

$$d\vec{r} = \frac{\partial \vec{r}}{\partial \xi_j} d\xi_j = \vec{e}_j d\xi_j = \hat{1}_i dx_i \tag{3.13}$$

In accord with Equation (3.10), the $\vec{e}_j$ basis is given by

$$\vec{e}_1 = h_1 \hat{t}, \quad \vec{e}_2 = h_2 \hat{n}, \quad \vec{e}_3 = h_3 \hat{b} \tag{3.14}$$

Conventional tensor notation would write ξ_j as ξ^j, because $\vec{e}_j$ is the basis for the ξ^j. In the interest of notational simplicity, this has not been done.

The direction cosines between the two bases are

$$a_{ij} = \hat{1}_i \cdot \hat{e}_j \tag{3.15}$$

which can be written as an array

$$a_{11} = \frac{1}{\left(1+f'^2\right)^{1/2}} \qquad a_{12} = \frac{f'}{\left(1+f'^2\right)^{1/2}} \qquad a_{13} = 0$$

$$a_{21} = \frac{x_2 f'}{R\left(1+f'^2\right)^{1/2}} \qquad a_{22} = -\frac{x_2}{R\left(1+f'^2\right)^{1/2}} \qquad a_{23} = \frac{\sigma x_3}{R} \qquad (3.16)$$

$$a_{31} = \frac{\sigma x_3 f'}{R\left(1+f'^2\right)^{1/2}} \qquad a_{32} = -\frac{\sigma x_3}{R\left(1+f'^2\right)^{1/2}} \qquad a_{33} = -\frac{x_2}{R}$$

They are functions of the Cartesian coordinates and $f' = df/dx_1$. The determinant of the a_{ij} is +1, which is the expected result for two orthonormal, right-handed bases.

The rightmost of Equation (3.13) is multiplied with $\cdot \vec{e}_j$, with the result

$$d\xi_j = \frac{a_{ij}}{h_j} dx_i, \quad \text{(no } j \text{ sum)}$$

This is compared with

$$d\xi_j = \frac{\partial \xi_j}{\partial x_i} dx_i$$

to obtain

$$\frac{\partial \xi_j}{\partial x_i} = \frac{a_{ij}}{h_j}, \quad \text{(no } j \text{ sum)} \qquad (3.17)$$

This relation is a key result and the reason for the a_{ij} equations. It is used to obtain the h_j and ξ_j.

Equation (3.12) yields for ξ_2

$$\frac{\partial \xi_2}{\partial x_1} = \frac{df}{dx_1} = f'$$

while Equation (3.17) results in

$$\frac{\partial \xi_2}{\partial x_1} = \frac{a_{12}}{h_2} = \frac{f'}{h_2\left(1+f'^2\right)^{1/2}}$$

We thus have

$$h_2 = \frac{1}{\left(1+f'^2\right)^{1/2}} \qquad (3.18)$$

Thus, only $j = 1$ and 3 need further consideration. Integration of Equation (3.17) for these two h_j values then yields an explicit form for the transformation equations. This integration, however, first requires evaluating h_1 and h_3 in terms of the x_i.

3.3   SCALE FACTORS

The scale factors are not arbitrary. They are established by the requirement that Equation (3.17) be integrable. This is assured if the change in the order of the differentiation compatibility condition (Stoker, 1969)

$$\frac{\partial^2 \xi_j}{\partial x_k \partial x_m} = \frac{\partial^2 \xi_j}{\partial x_m \partial x_k}, \quad j = 1,2,3, \; m \neq k \tag{3.19}$$

is satisfied. (This condition is not always satisfied as demonstrated in Appendix K.) Since ξ_2 is given by Equation (3.12), it satisfies the compatibility condition and only $j = 1, 3$ need be considered. For each j, this equation represents three equations. In combination with Equation (3.17), these become

$$a_{mj}\frac{\partial q_j}{\partial x_k} - a_{kj}\frac{\partial q_j}{\partial x_m} = \frac{\partial a_{mj}}{\partial x_k} - \frac{\partial a_{kj}}{\partial x_m}$$

where $q_j = \ell n h_j$. When written out, we have

$$a_{2j}\frac{\partial q_j}{\partial x_1} - a_{1j}\frac{\partial q_j}{\partial x_2} = \frac{\partial a_{2j}}{\partial x_1} - \frac{\partial a_{1j}}{\partial x_2} \tag{3.20a}$$

$$a_{3j}\frac{\partial q_j}{\partial x_1} - a_{1j}\frac{\partial q_j}{\partial x_3} = \frac{\partial a_{3j}}{\partial x_1} - \frac{\partial a_{1j}}{\partial x_3} \tag{3.20b}$$

$$a_{3j}\frac{\partial q_j}{\partial x_2} - a_{2j}\frac{\partial q_j}{\partial x_3} = \frac{\partial a_{3j}}{\partial x_2} - \frac{\partial a_{2j}}{\partial x_3} \tag{3.20c}$$

for $j = 1$ and 3. With $\partial q_j/\partial x_i$ as unknowns, the value of the determinant of the left side is zero. Hence, elimination of the q_j derivatives results in a condition on the a_{ij} coefficients:

$$a_{2j}^2\frac{\partial}{\partial x_1}\left(\frac{a_{3j}}{a_{2j}}\right) + a_{3j}^2\frac{\partial}{\partial x_2}\left(\frac{a_{1j}}{a_{3j}}\right) + a_{1j}^2\frac{\partial}{\partial x_3}\left(\frac{a_{2j}}{a_{1j}}\right) = 0, \quad j = 1,3 \tag{3.21}$$

for the existence of a solution of Equation (3.20). This equation can be shown to hold for all j, including $j = 2$ (see Problem 3). Thus, a solution of Equation (3.20) exists for the h_j scale factors of a two-dimensional or axisymmetric shock.

Each of the equations in Equation (3.20) is a separate equation for q_j and is solved independently of the other two. (The system of equations is overdetermined.) Each of these solutions involves an arbitrary function of integration. There are no boundary or initial conditions that can be used to evaluate these functions of integration. Instead, they are chosen in order that the resulting q_j is a solution of all three of the equations in Equation (3.20). Superscripts a, b, and c, respectively, are used to denote the solutions of these equations. They are first-order partial differential equations (PDEs) and their general solution is obtained by the method-of-characteristics (MOC) in Appendix C. With this approach, we obtain from Equation (3.20a) the characteristic equations (Equation C.8 in Appendix C):

$$\frac{dx_1}{a_{2j}} = -\frac{dx_2}{a_{1j}} = \frac{dx_3}{0} = \frac{dq_j^{(a)}}{\dfrac{\partial a_{2j}}{\partial x_1} - \dfrac{\partial a_{1j}}{\partial x_2}} \tag{3.22}$$

For a fixed j, let $u_{jk}^{(a)}(x_i)$, $k = 1,2,3$, denote the functional form of the unique solutions to these three first-order ordinary differential equations (ODEs), where $u_{jk}^{(a)} = c_{jk}^{(a)}$ and $c_{jk}^{(a)}$ are the integration constants (Equation C.7). To avoid an infinity, the dx_3 term is made indeterminant by setting x_3 equal to a constant. The solutions of the two leftmost ODEs are written as

$$u_{j1}^{(a)} = x_3 = c_{j1}^{(a)}, \quad u_{j2}^{(a)} = c_{j2}^{(a)} \tag{3.23a,b}$$

The relation $u_{j2}^{(a)} = c_{j2}^{(a)}$ is the functional form for the solution of the leftmost of Equation (3.22); x_3 is held fixed in a_{1j} and a_{2j} when obtaining this solution. There are two equivalent possibilities for $q_j^{(a)}$; for purposes of brevity only one is presented. The functional form for the solution of the dx_1, $dq_j^{(a)}$ equation is written as

$$u_{j3}^{(a)}\left(x_i, q_j^{(a)}\right) = q_j^{(a)} - \int \left(\frac{\partial a_{2j}}{\partial x_1} - \frac{\partial a_{1j}}{\partial x_2}\right) \frac{dx_1}{a_{2j}} = c_{j3}^{(a)} \tag{3.23c}$$

where, if necessary, x_2 and x_3 are replaced in the integrand with the aid of Equation (3.23a,b). After the integration is performed, the constants $c_{jk}^{(a)}$, $k = 1,2$ are then replaced by x_3, which equals $u_{j1}^{(a)}$, and by $u_{j2}^{(a)}$.

Although theoretically equivalent, the quadrature that results from using dx_2, $dq_j^{(a)}$ may be simpler or more complicated than the one stemming from dx_1, $dq_j^{(a)}$. In either case, the general solution of Equation (3.20a) is rewritten as

$$u_{j3}^{(a)} = \ell n g_j^{(a)}\left(u_{j1}^{(a)}, u_{j2}^{(a)}\right)$$

in accordance with Equation (C.9), where $g_j^{(a)}$ is an arbitrary function of its two arguments. Equation (3.23c) with $q_j^{(a)} = \ell n h_j^{(a)}$ and $u_{j3}^{(a)}$ replaced with $\ell n g_j^{(a)}$ then yields

$$\ell n h_j^{(a)} = \int \left(\frac{\partial a_{2j}}{\partial x_1} - \frac{\partial a_{1j}}{\partial x_2}\right) \frac{dx_1}{a_{2j}} + \ell n g_j^{(a)}\left[x_3, u_{j2}^{(a)}(x_i)\right]$$

or finally

$$h_j^{(a)} = g_j^{(a)}\left(x_3, u_{j2}^{(a)}\right)\exp\left[\int\left(\frac{\partial a_{2j}}{\partial x_1} - \frac{\partial a_{1j}}{\partial x_2}\right)\frac{dx_1}{a_{2j}}\right] \tag{3.24a}$$

The same procedure, when applied to Equation (3.20b,c), results in

$$h_j^{(b)} = g_j^{(b)}\left(x_2, u_{j2}^{(b)}\right)\exp\left[\int\left(\frac{\partial a_{3j}}{\partial x_1} - \frac{\partial a_{1j}}{\partial x_3}\right)\frac{dx_1}{a_{3j}}\right] \tag{3.24b}$$

$$h_j^{(c)} = g_j^{(c)}\left(x_1, u_{j2}^{(c)}\right)\exp\left[-\int\left(\frac{\partial a_{3j}}{\partial x_2} - \frac{\partial a_{2j}}{\partial x_3}\right)\frac{dx_3}{a_{2j}}\right] \tag{3.24c}$$

where $u_{j2}^{(b)}$ and $u_{j2}^{(c)}$ are solutions of

$$\frac{dx_1}{a_{3j}} = -\frac{dx_3}{a_{1j}}, \quad \frac{dx_2}{a_{3j}} = -\frac{dx_3}{a_{2j}}$$

respectively. The various g_j coefficients are chosen by inspection so that

$$h_j = h_j^{(a)} = h_j^{(b)} = h_j^{(c)}, \quad j = 1,3 \tag{3.25}$$

In view of this constraint, the g_j selection must satisfy

$$g_j^{(a)}\exp\left[\int\left(\frac{\partial a_{2j}}{\partial x_1} - \frac{\partial a_{1j}}{\partial x_2}\right)\frac{dx_1}{a_{2j}}\right] = g_j^{(b)}\exp\left[\int\left(\frac{\partial a_{3j}}{\partial x_1} - \frac{\partial a_{1j}}{\partial x_3}\right)\frac{dx_1}{a_{3j}}\right]$$

$$= g_j^{(c)}\exp\left[-\int\left(\frac{\partial a_{3j}}{\partial x_2} - \frac{\partial a_{2j}}{\partial x_3}\right)\frac{dx_3}{a_{2j}}\right]$$

3.4 APPLICATION TO A TWO-DIMENSIONAL OR AXISYMMETRIC SHOCK

Equation (3.20a) is developed for $j = 1$:

$$a_{21}\frac{\partial q_1^{(a)}}{\partial x_1} - a_{11}\frac{\partial q_1^{(a)}}{\partial x_2} = \frac{\partial a_{21}}{\partial x_1} - \frac{\partial a_{11}}{\partial x_2} \tag{3.26a}$$

where

$$\frac{\partial a_{11}}{\partial x_2} = 0$$

$$\frac{\partial a_{21}}{\partial x_1} = \frac{x_2}{R} \frac{d}{dx_1}\left[\frac{f'}{\left(1+f'^2\right)^{1/2}}\right] = \frac{x_2 f''}{R\left(1+f'^2\right)^{3/2}}$$

Equation (3.26a) simplifies to

$$\frac{x_2}{R} f' \frac{\partial q_1^{(a)}}{\partial x_1} - \frac{\partial q_1^{(a)}}{\partial x_2} - \frac{x_2}{R} \frac{f''}{\left(1+f'^2\right)} = 0 \qquad (3.26b)$$

The characteristic equations are

$$\frac{R}{x_2} \frac{dx_1}{f'} = -dx_2 = \frac{dx_3}{0} = \frac{R}{x_2} \frac{1+f'^2}{f''} dq_1^{(a)}$$

which yield

$$x_3 = c_1$$

$$\frac{dx_1}{f'} + \frac{x_2 dx_2}{\left(x_2^2 + \sigma c_1^2\right)^{1/2}} = 0$$

$$\frac{dq_1^{(a)}}{dx_1} - \frac{1}{f'\left(1+f'^2\right)} \frac{df'}{dx_1} = 0$$

The second of these equations has the integral

$$\int \frac{x_2 dx_2}{\left(x_2^2 + \sigma c_1^2\right)^{1/2}} = \left(x_2^2 + \sigma c_1^2\right)^{1/2} = \left(x_2^2 + \sigma x_3^2\right)^{1/2} = R$$

with the result

$$R + \int \frac{dx_1}{f'} = c_2$$

The third characteristic equation has the integral

$$\int \frac{df'}{f'\left(1+f'^2\right)} = \ell n \frac{f'}{\left(1+f'^2\right)^{1/2}}$$

and

$$q_1^{(a)} - \ell n \frac{f'}{\left(1+f'^2\right)^{1/2}} = c_3$$

where the integration constants, c_k, equal $u_{1k}^{(a)}$. We thus obtain

$$q_1^{(a)} = \ell n h_1^{(a)} = \ell n \frac{f'}{\left(1+f'^2\right)^{1/2}} + \ell n g_1^{(a)}\left(x_3, R + \int \frac{dx_1}{f'}\right)$$

or

$$h_1^{(a)} = \frac{f'}{\left(1+f'^2\right)^{1/2}} g_1^{(a)}\left(x_3, R + \int \frac{dx_1}{f'}\right)$$

where $g_1^{(a)}$ is an arbitrary function of its two arguments.
   The same process for Equations (3.20b) and (3.20c) yields

$$h_1^{(b)} = \frac{f'}{\left(1+f'^2\right)^{1/2}} g_1^{(b)}\left(x_2, R + \int \frac{dx_1}{f'}\right)$$

$$h_1^{(c)} = g_1^{(c)}\left(x_1, R\right)$$

A simple choice is

$$g_1^{(a)} = 1, \quad g_1^{(b)} = 1, \quad g_1^{(c)} = \frac{f'}{\left(1+f'^2\right)^{1/2}}$$

because f' is a function only of x_1. Consequently, h_1 is

$$h_1 = \frac{f'}{\left(1+f'^2\right)^{1/2}} \tag{3.27}$$

Equation (3.20a,b) for $j = 3$ yields

$$q_3^{(a)} = q_3^{(b)} = g_3\left(x_2, x_3\right)$$

while Equation (3.20c) reduces to

$$\frac{x_2}{R}\frac{\partial q_3^{(c)}}{\partial x_2} + \frac{\sigma x_3}{R}\frac{\partial q_3^{(c)}}{\partial x_3} = \frac{\sigma}{R}$$

For $\sigma = 0$, it is evident that $h_3 = 1$. When $\sigma = 1$, it is simpler to forego the MOC and simply assume

$$q_3^{(c)} = g(R)$$

By substituting this into the above PDE, we readily obtain

$$q_3^{(c)} = \ell nR, \quad \sigma = 1$$

The scale factors now summarize as

$$h_1 = \frac{f'}{\left(1+f'^2\right)^{1/2}}, \quad h_2 = \frac{1}{\left(1+f'^2\right)^{1/2}}, \quad h_3 = R^\sigma \qquad (3.28)$$

3.5   TRANSFORMATION EQUATIONS

Our next task is to use Equation (3.17) to obtain the transformation equations. For $j = 1$, we write

$$\frac{\partial \xi_1}{\partial x_1} = \frac{a_{11}}{h_1} = \frac{1}{f'}$$

which integrates to

$$\xi_1 = \varphi_1(x_2, x_3) + \int \frac{dx_1}{f'}$$

where φ_1 is a function of integration. To evaluate this function, we use

$$\frac{\partial \xi_1}{\partial x_2} = \frac{\partial \varphi_1}{\partial x_2}$$

$$\frac{\partial \xi_1}{\partial x_2} = \frac{a_{21}}{h_1} = \frac{x_2}{R}$$

or

$$\frac{\partial \varphi_1}{\partial x_2} = \frac{x_2}{R}$$

$$\varphi_1 = \varphi_2(x_3) + \int \frac{x_2 dx_2}{\left(x_2^2 + \sigma x_3^2\right)^{1/2}}$$

In the integrand, x_3 is held constant, and φ_2 is a second function of integration. Evaluation of the integral yields

$$\varphi_1 = \varphi_2 + \left(x_2^2 + \sigma x_3^2\right)^{1/2} = \varphi_2 + R$$

To evaluate φ_2, we utilize

$$\frac{\partial \xi_1}{\partial x_3} = \frac{d\varphi_2}{dx_3} + \frac{\sigma x_3}{R} = \frac{a_{32}}{h_3} = \frac{\sigma x_3}{R}$$

As a consequence, φ_2 is given by

$$\frac{d\varphi_2}{dx_3} = 0$$

or

$$\varphi_2 = 0$$

Thus, the ξ_1 transformation equation is

$$\xi_1 = R + \int_0^{x_1} \frac{dx_1}{f'}$$

For ξ_3, we use

$$\frac{\partial \xi_3}{\partial x_i} = \frac{a_{i3}}{h_3} = \frac{a_{i3}}{R^\sigma}$$

or

$$\frac{\partial \xi_3}{\partial x_1} = 0, \quad \frac{\partial \xi_3}{\partial x_2} = \frac{\sigma x_3}{R^2}, \quad \frac{\partial \xi_3}{\partial x_3} = -\frac{x_2}{R^{1+\sigma}}$$

When $\sigma = 0$, with $R = x_2$, this system provides

$$\xi_3 = -x_3, \quad \sigma = 0$$

as expected. When $\sigma = 1$, direct integration yields

$$\xi_3 = \tan^{-1}\left(\frac{x_2}{x_3}\right), \quad \sigma = 1$$

This result is readily checked by differentiation.

With the aid of Equation (3.12), the transformation equations are

$$\xi_1 = R + \int_0^{x_1} \frac{dx_1}{f'}, \quad \xi_2 = f - R, \quad \xi_3 = (\sigma - 1)x_3 + \sigma \tan^{-1}\left(\frac{x_2}{x_3}\right) \tag{3.29}$$

The scale factor and transformation equations are an explicit result for any differentiable two-dimensional or axisymmetric shock. They provide a global orthogonal coordinate system where $\xi_2 = 0$ on the surface with ξ_1 and ξ_3 as surface coordinates. As various checks, one can show that the equations in Equation (3.20) are satisfied for $j = 1, 2, 3$, the $\hat{e}_i$ orthonormal basis is tangent to the ξ_i, and the arc length is

$$(ds)^2 = h_j^2 (d\xi_j)^2 = (dx_1)^2 + (dx_2)^2 + (dx_3)^2 \tag{3.30}$$

Problems 4 and 5 further illustrate the application of the foregoing analysis.

3.6 BASIS DERIVATIVES

It is simpler to use (Appendix D)

$$\frac{\partial \hat{e}_j}{\partial \xi_i} = \sum_{k \neq j} \left(\frac{\delta_{ik}}{h_j} \frac{\partial h_i}{\partial \xi_j} - \frac{\delta_{ij}}{h_k} \frac{\partial h_j}{\partial \xi_k} \right) \hat{e}_k \tag{3.31}$$

for orthogonal coordinates rather than the more general Christoffel symbol for the $\hat{e}_j$ derivatives. For this, the h_i must be functions of ξ_j rather than functions of x_j.

This is neatly accomplished with Jacobian theory (Emanuel 2001, Appendix B). The following two arrays of derivatives with respect to the x_i are utilized:

$$\frac{\partial h_1}{\partial x_1} = \frac{f''}{\left(1+f'^2\right)^{3/2}}, \quad \frac{\partial h_1}{\partial x_2} = 0, \quad \frac{\partial h_1}{\partial x_3} = 0$$

$$\frac{\partial h_2}{\partial x_1} = -\frac{f'f''}{\left(1+f'^2\right)^{3/2}}, \quad \frac{\partial h_2}{\partial x_2} = 0, \quad \frac{\partial h_2}{\partial x_3} = 0 \tag{3.32}$$

$$\frac{\partial h_3}{\partial x_1} = 0, \quad\quad\quad \frac{\partial h_3}{\partial x_2} = \frac{\sigma x_2}{R}, \quad \frac{\partial h_3}{\partial x_3} = \frac{\sigma x_3}{R}$$

and

$$\frac{\partial \xi_1}{\partial x_1} = \frac{1}{f'}, \quad \frac{\partial \xi_1}{\partial x_2} = \frac{x_2}{R}, \quad \frac{\partial \xi_1}{\partial x_3} = \frac{\sigma x_3}{R}$$

$$\frac{\partial \xi_2}{\partial x_1} = f', \quad \frac{\partial \xi_2}{\partial x_2} = -\frac{x_2}{R}, \quad \frac{\partial \xi_2}{\partial x_3} = -\frac{\sigma x_3}{R} \tag{3.33}$$

$$\frac{\partial \xi_3}{\partial x_1} = 0, \quad \frac{\partial \xi_3}{\partial x_2} = \frac{\sigma x_3}{R^2}, \quad \frac{\partial \xi_3}{\partial x_3} = -\left(\frac{x_2}{R^2}\right)^{\sigma}$$

The Jacobian of the transformation reduces to

$$J = \frac{\partial(\xi_1, \xi_2, \xi_3)}{\partial(x_1, x_2, x_3)} = \frac{1+f'^2}{f'} \frac{1}{R^{\sigma}} \tag{3.34}$$

As an illustration, the $\partial h_1/\partial \xi_1$ derivative is given by

$$\frac{\partial h_1}{\partial \xi_1} = \frac{\partial(h_1, \xi_2, \xi_3)}{\partial(\xi_1, \xi_2, \xi_3)} = \frac{\dfrac{\partial(h_1, \xi_2, \xi_3)}{\partial(x_1, x_2, x_3)}}{\dfrac{\partial(\xi_1, \xi_2, \xi_3)}{\partial(x_1, x_2, x_3)}}$$

$$= \frac{1}{J} \begin{vmatrix} \dfrac{\partial h_1}{\partial x_1} & 0 & 0 \\[2mm] \dfrac{\partial \xi_2}{\partial x_1} & \dfrac{\partial \xi_2}{\partial x_2} & \dfrac{\partial \xi_2}{\partial x_3} \\[2mm] \dfrac{\partial \xi_3}{\partial x_1} & \dfrac{\partial \xi_3}{\partial x_2} & \dfrac{\partial \xi_3}{\partial x_3} \end{vmatrix} = \frac{1}{J}\frac{\partial h_1}{\partial x_1}\left(\frac{\partial \xi_2}{\partial x_2}\frac{\partial \xi_3}{\partial x_3} - \frac{\partial \xi_2}{\partial x_3}\frac{\partial \xi_3}{\partial x_2}\right) = \frac{f'f''}{\left(1+f'^2\right)^{5/2}},$$

The h_i derivates used in Equation (3.31) are then

$$\frac{\partial h_1}{\partial \xi_1} = \frac{f' f''}{\left(1 + f'^2\right)^{5/2}}, \quad \frac{\partial h_1}{\partial \xi_2} = \frac{f' f''}{\left(1 + f'^2\right)^{5/2}}, \quad \frac{\partial h_1}{\partial \xi_3} = 0$$

$$\frac{\partial h_2}{\partial \xi_1} = -\frac{f'^2 f''}{\left(1 + f'^2\right)^{5/2}}, \quad \frac{\partial h_2}{\partial \xi_2} = -\frac{f'^2 f''}{\left(1 + f'^2\right)^{5/2}}, \quad \frac{\partial h_2}{\partial \xi_3} = 0 \qquad (3.35)$$

$$\frac{\partial h_3}{\partial \xi_1} = \frac{\sigma f'^2}{\left(1 + f'^2\right)}, \quad \frac{\partial h_3}{\partial \xi_2} = -\frac{\sigma}{\left(1 + f'^2\right)}, \quad \frac{\partial h_3}{\partial \xi_3} = 0$$

Equations (3.28), (3.31), and (3.35) now yield

$$\frac{\partial \hat{e}_1}{\partial \xi_1} = -\frac{f' f''}{\left(1 + f'^2\right)^2} \hat{e}_2, \quad \frac{\partial \hat{e}_1}{\partial \xi_2} = -\frac{f' f''}{\left(1 + f'^2\right)^2} \hat{e}_2, \quad \frac{\partial \hat{e}_1}{\partial \xi_3} = \frac{\sigma f'}{\left(1 + f'^2\right)^{1/2}} \hat{e}_3$$

$$\frac{\partial \hat{e}_2}{\partial \xi_1} = \frac{f' f''}{\left(1 + f'^2\right)^2} \hat{e}_1, \quad \frac{\partial \hat{e}_2}{\partial \xi_2} = \frac{f' f''}{\left(1 + f'^2\right)^2} \hat{e}_2, \quad \frac{\partial \hat{e}_2}{\partial \xi_3} = -\frac{\sigma}{\left(1 + f'^2\right)^{1/2}} \hat{e}_3 \qquad (3.36)$$

$$\frac{\partial \hat{e}_3}{\partial \xi_1} = 0, \quad \frac{\partial \hat{e}_3}{\partial \xi_2} = 0, \quad \frac{\partial \hat{e}_3}{\partial \xi_3} = -\frac{\sigma}{\left(1 + f'^2\right)^{1/2}} \left(f' \hat{e}_1 - \hat{e}_2\right)$$

For the subsequent analysis, it is analytically convenient to utilize the $\hat{t}, \hat{n}, \hat{b}$ basis, the s, n, b coordinates, and β. Equation (3.9) and

$$\partial s = h_1 \partial \xi_1, \quad \partial n = h_2 \partial \xi_2, \quad \partial b = h_3 \partial \xi_3 \qquad (3.37)$$

are utilized. To replace f' and f'', we use

$$f' = \tan \beta, \quad \left(1 + f'^2\right)^{1/2} = \frac{1}{\cos \beta} \qquad (3.38a,b)$$

and

$$s = \int_0^{x_1} \left[1 + \left(\frac{dR}{dx_1} \right)^2 \right]^{1/2} dx_1 = \int_0^{x_1} \left(1 + f'^2\right)^{1/2} dx_1 = \int_0^{x_1} \frac{dx_1}{\cos \beta} \qquad (3.39)$$

or

$$\frac{ds}{dx_1} = \frac{1}{\cos \beta} \qquad (3.40)$$

It is convenient to introduce

$$\beta' = \frac{d\beta}{ds} = \frac{d\beta}{dx_1}\frac{dx_1}{ds} = \frac{d\beta}{dx_1}\cos\beta \tag{3.41}$$

where $-\beta'$ is the flow plane's curvature of the shock, as suggested by Figure 2.1. Note that the primes on f and β are with respect to x_1 and s, respectively. From Equation (3.38a), we have

$$f'' = \frac{d^2 f}{dx_1^2} = \frac{d(\tan\beta)}{dx_1} = \frac{1}{\cos^2\beta}\frac{d\beta}{dx_1}$$

Eliminate $d\beta/dx_1$ from this equation and Equation (3.41), with the result

$$f'' = \frac{\beta'}{\cos^3\beta} \tag{3.42}$$

Equations (3.38a) and (3.42) are used to replace f' and f''.

   With the foregoing, Equation (3.36) becomes

$$\frac{\partial \hat{t}}{\partial s} = -\beta'\,\hat{n},\quad \frac{\partial \hat{t}}{\partial n} = -\beta'\tan\beta\hat{n},\quad \frac{\partial \hat{t}}{\partial b} = \frac{\sigma\sin\beta}{R}\hat{b}$$

$$\frac{\partial \hat{n}}{\partial s} = \beta'\hat{t},\quad \frac{\partial \hat{n}}{\partial n} = \beta'\tan\beta\hat{t},\quad \frac{\partial \hat{n}}{\partial b} = -\frac{\sigma\cos\beta}{R}\hat{b} \tag{3.43}$$

$$\frac{\partial \hat{b}}{\partial s} = 0,\quad \frac{\partial \hat{b}}{\partial n} = 0,\quad \frac{\partial \hat{b}}{\partial b} = \frac{\sigma}{R}\left(-\sin\beta\hat{t} + \cos\beta\hat{n}\right)$$

where R is still given by Equation (3.1b). As expected, derivatives with respect to b are zero in a two-dimensional flow. The gradient operator, Equation (3.9), now becomes

$$\nabla = \hat{t}\frac{\partial}{\partial s} + \hat{n}\frac{\partial}{\partial n} + \hat{b}\frac{\partial}{\partial b} \tag{3.44}$$

4 Derivatives for a Two-Dimensional or Axisymmetric Shock with a Uniform Freestream

4.1 PRELIMINARY REMARKS

In the next section, a concise notation is introduced that is used throughout the rest of the manuscript. It is especially convenient for shock wave studies. All equations are dimensional, where the shock shape, Equation (3.2), γ, the upstream pressure, p_1, density, ρ_1, and flow speed, V_1, or M_1, are presumed to be known. These parameters are sufficient for nondimensionalizing the equations. Thus, results, summarized in Appendix E, are in terms of these quantities. The appendix should be useful for the development of computational algorithms.

In the tangential and normal derivative equations, the quantities

$$1/\beta', s, n, y(= R)$$

have dimensions of length. Because they appear homogeneously in the equations, they may be dimensional or nondimensional. In fact, the steady Euler equations, in any coordinate system, are homogeneous with respect to an inverse length. Because the jump conditions are independent of length, a normalizing length is arbitrary.

As noted in Chapter 1, the jump conditions hold for an unsteady, three-dimensional shock. Both tangential and normal derivatives are in the flow plane, as is the shock's curvature, $-\beta'$. On the other hand, the normal derivatives stem from a derivation that requires the steady Euler equations. In these equations, continuity contains the dimensionality parameter, σ. The normal derivatives are therefore constrained to a steady, two-dimensional or axisymmetric shock.

4.2 JUMP CONDITIONS

Equations (2.22) through (2.25) become

$$V_2 = V_1 \frac{\cos\beta}{\cos(\beta-\theta)} \tag{4.1a}$$

$$\rho_2 = \rho_1 \frac{\tan\beta}{\tan(\beta-\theta)} \tag{4.1b}$$

$$p_2 = p_1 + (\rho V^2)_1 \frac{\sin\beta\sin\theta}{\cos(\beta-\theta)} \tag{4.1c}$$

$$h_2 = h_1 + \frac{1}{2}V_1^2 \frac{\sin\theta\sin(2\beta-\theta)}{\cos^2(\beta-\theta)} \tag{4.1d}$$

for the jump conditions.

The perfect gas assumption and Mach numbers provided by Equations (2.34a,b) are introduced. The notation is simplified by defining

$$m = M_1^2, \quad w = (M_1\sin\beta)^2 \tag{4.2}$$

$$X = 1 + \frac{\gamma-1}{2}w, \; Y = \gamma w - \frac{\gamma-1}{2}, \; Z = w-1, \; A = \frac{\gamma+1}{2}\frac{m\sin\beta\cos\beta}{X}, \; B = 1+A^2$$

Note that $M_1\sin\beta$ is the normal component of M_1.

For the analysis, it is convenient to use velocity components u and v that are tangential and normal to the shock, respectively, as sketched in Figure 3.1. These components are related to V_1 and V_2 by means of (see Figure 2.1 for β and θ)

$$u_1 = u_2 = V_1\cos\beta = V_2\cos(\beta-\theta) \tag{4.3a}$$

$$v_1 = V_1\sin\beta, \quad v_2 = V_2\sin(\beta-\theta) \tag{4.3b}$$

Equation (4.1), in combination with perfect gas thermodynamic state equations, then yield the jump conditions in terms of u and v. These equations are summarized in Appendix E.1 in Appendix E, which shows several Mach number functions, because these appear in later equations. Equations for $\tan\theta$, $\sin\theta$, $\sin(\beta-\theta)$, and $\cos(\beta-\theta)$ are also listed. The equation for $\tan\theta$ easily reduces to Equation (2.28), while $\sin\theta$ is the subject of Problem 6.

The equations in Appendix E.1 are arrived at by replacing the normal component of the Mach numbers, M_{1n} and M_{2n}, with

$$M_{1n} = M_1\sin\beta, \quad M_{2n} = M_2\sin(\beta-\theta) \tag{4.4}$$

in the standard equations for a normal shock. The results appear different because $\sin(\beta-\theta)$ has been systematically eliminated. For instance, the usual equation for M_2 can be written as

$$M_2^2 = \frac{1}{\sin^2(\beta-\theta)}\frac{1+\frac{\gamma-1}{2}M_1^2\sin^2\beta}{\gamma M_1^2\sin^2\beta - \frac{\gamma-1}{2}} = \frac{1}{\sin^2(\beta-\theta)}\frac{X}{Y}$$

With the $\sin(\beta-\theta)$ relation in the appendix, the listed M_2^2 equation is obtained.

4.3   TANGENTIAL DERIVATIVES

The equations in Appendix E.1 are differentiated with respect to the arc length, s, along the shock in the flow plane. The resulting derivatives are proportional to β', where

$$\beta' = \frac{d\beta}{ds} = \frac{f''}{\left(1 + f'^2\right)^{3/2}} \tag{4.5}$$

and $-\beta'$ is the shock's curvature in the flow plane. As previously noted, this curvature is positive when the shock is convex relative to the upstream flow and β' is negative. It is useful for the latter discussion to include the derivative of the Mach angle μ:

$$\mu = \sin^{-1}\frac{1}{M} \tag{4.6}$$

and the included angle θ between $\bar{V}_1$ and $\bar{V}_2$.

To illustrate how Appendix E.2 is obtained, the derivative of the stagnation pressure is obtained, starting with $p_{o,2}$ in Appendix E.1:

$$
\begin{aligned}
\frac{1}{p_1}\left(\frac{\partial p_o}{\partial s}\right)_2 &= \frac{2}{(\gamma+1)}\left\{\left(1+\frac{\gamma-1}{2}M_2^2\right)^{\gamma/(\gamma-1)}\frac{\partial Y}{\partial s}\right. \\
&\qquad \left. +\frac{\gamma}{\gamma-1}Y\left(1+\frac{\gamma-1}{2}M_2^2\right)^{[\gamma/(\gamma-1)]-1}\frac{\gamma-1}{2}\left(\frac{\partial M^2}{\partial s}\right)_2\right\} \\[2mm]
&= \frac{2}{(\gamma+1)}\left(1+\frac{\gamma-1}{2}M_2^2\right)^{\gamma/(\gamma-1)}\left[2\gamma\ m\beta'\sin\beta\cos\beta\right. \\[2mm]
&\qquad \left. -\frac{\gamma}{2}\frac{Y}{1+\frac{\gamma-1}{2}M_2^2}\frac{(\gamma+1)^2}{2}\left(1+\frac{\gamma-1}{2}m\right)\left(1+\gamma w^2\right)\frac{\beta'm\sin\beta\cos\beta}{X^2Y^2}\right] \\[2mm]
&= \frac{2\gamma}{\gamma+1}\left(1+\frac{\gamma-1}{2}M_2^2\right)^{\gamma/(\gamma-1)}\beta'm\sin\beta\cos\beta\left(2-\frac{1+\gamma w^2}{wX}\right) \\[2mm]
&= -\frac{2\gamma}{\gamma+1}\left(1+\frac{\gamma-1}{2}M_2^2\right)^{\gamma/(\gamma-1)}\frac{\beta'\ Z^2}{X\ \tan\beta} \tag{4.7}
\end{aligned}
$$

Observe that $(\partial M^2/\partial s)_2$ is used in the derivation, but it is inconvenient to replace the factor containing M_2^2.

Although σ does not appear in Appendices E.1 and E.2, all results hold for an axisymmetric shock. Except for $(\partial u/\partial s)_2$ and $(\partial\theta/\partial s)_2$, the listed derivatives are proportional to $\cos\beta$, which means they are zero when the shock is normal to the freestream velocity.

4.4 NORMAL DERIVATIVES

The steady Euler equations are needed in a scalar form and with orthogonal coordinates, where one coordinate is along the shock in the flow plane and the other is normal to it. Emanuel (1986, Section 13.3) derives these equations in this form, and Table 4.1 provides the change to our notation. The minus sign that appears with ∂n and v stems from the downstream orientation of the n coordinate. Note the replacement of x_2 with y, which also replaces the R of Equations (3.2) and (3.12). The angle θ in Emanuel (1986) is the angle of the ξ_1 coordinate with respect to the x_1-axis. When the coordinate system is rotated to align it with the shock, ξ_1 becomes the s coordinate, and θ becomes β. (Remember that $\bar{V}_1$ is parallel to x_1.) The version of the Euler equations, given shortly, applies only to the flow field just downstream of a shock.

The analysis in the above reference is for an arbitrary point in a steady, two-dimensional or axisymmetric flow. Here, the equations are written for a point just downstream of a shock. Consequently, κ_1 is the longitudinal curvature of the shock. The κ_o curvature is for the n-coordinate in the flow plane. The above reference shows that

$$\kappa_s = -\frac{\partial\theta}{\partial s}, \quad \kappa_o = \frac{\partial\theta}{\partial n} \tag{4.8}$$

where on the shock surface, θ now becomes β, and β is a function only of s. We thus obtain

$$\kappa_s = -\frac{d\beta}{ds} = -\beta', \quad \kappa_o = 0 \tag{4.9a,b}$$

In general, κ_o is not zero. It is zero here because the analysis is restricted to a surface where n is a constant, which is in accord with Hayes (1957). The κ_o parameter appears in continuity and the two flow plane scalar momentum equations in

---

TABLE 4.1
Transformation to the Current Notation

Emanuel (1986)	Present Notation
$h_1\partial\xi_1$	∂s
$h_2\partial\xi_2$	$-\partial n$
x_2	y
v_1	u
v_2	$-v$
κ_1	$-\beta'$
κ_2	0
θ	β

Emanuel (1986). These three terms containing κ_o have been deleted from Equation (4.10). Problem 7, for the s and n momentum equations, demonstrates that the deletion is required for consistency with the formulas for $(\partial p/\partial s)_2$ and $(\partial p/\partial n)_2$ in Appendix E.

A consequence of the normal derivative analysis is that the streamline derivative of the entropy, S, or stagnation pressure, p_o, should be zero. This streamline derivative, developed in Section 5.2, directly depends on the s and n derivatives. If there is an error in the normal derivative analysis, then the streamline derivatives of p_o and S would not be identically zero. Problem 8 shows this is not the case.

Problem 7 also demonstrates the κ_o being zero is actually a consequence, for a surface evaluation, that $\partial \hat{t}/\partial n$, $\partial \hat{n}/\partial n$, and $\partial \hat{b}/\partial n$ are zero. The first two derivatives are proportional to $\tan\beta$, where the $\tan\beta$ in $\partial \hat{n}/\partial n$ results in an infinity for $(\partial p/\partial s)_2$ at a normal shock, whereas the $\tan\beta$ in $\partial \hat{t}/\partial n$ does not (Problem 7). The zero for $\partial \hat{b}/\partial n$ stems directly from Equation (3.43).

For notational simplicity, the subscript 2 is suppressed that should appear on all variables and derivatives. This suppression, except for purposes of clarity, is used in the balance of this monograph. The Euler equations are written as

$$\frac{\partial(\rho u)}{\partial s}+\frac{\partial(\rho v)}{\partial n}+\beta'\rho v+\frac{\sigma\rho}{y}\left(u\sin\beta-v\cos\beta\right)=0$$

$$u\frac{\partial u}{\partial s}+v\frac{\partial u}{\partial n}+\beta'uv+\frac{1}{\rho}\frac{\partial p}{\partial s}=0$$

$$u\frac{\partial v}{\partial s}+v\frac{\partial v}{\partial n}-\beta'u^2+\frac{1}{\rho}\frac{\partial p}{\partial n}=0 \tag{4.10}$$

$$\left(u\frac{\partial}{\partial s}+v\frac{\partial}{\partial n}\right)\left[\frac{\gamma}{\gamma-1}\frac{p}{\rho}+\frac{1}{2}\left(u^2+v^2\right)\right]=0$$

The values of the u, v, p, and ρ variables and their s derivatives are known from Appendices E.1 and E.2. For instance, for the $\partial(\rho u)/\partial s$ term in continuity, we use

$$\frac{\partial(\rho u)}{\partial s}=\rho\frac{\partial u}{\partial s}+u\frac{\partial\rho}{\partial s}=\left(\frac{\gamma+1}{2}\rho_1\frac{w}{X}\right)\left(-V_1\beta'\sin\beta\right)$$

$$+\left(V_1\cos\beta\right)\left[(\gamma+1)\rho_1\frac{\beta'm\sin\beta\cos\beta}{X^2}\right]$$

$$=(\gamma+1)(\rho V)_1\frac{\beta'\sin\beta}{X^2}\left(-\frac{1}{2}wX+m\cos^2\beta\right)$$

$$=(\gamma+1)(\rho V)_1\frac{\beta'\sin\beta}{X^2}\left(m-\frac{3}{2}w-\frac{\gamma-1}{4}w^2\right)$$

The equations in Equation (4.10) are four, linear, inhomogeneous, algebraic equations for $\partial u/\partial n$, $\partial v/\partial n$, $\partial p/\partial n$, and $\partial \rho/\partial n$. The solution of these equations, obtained with the assistance of the MACSYMA code (Rand, 1984), is given in Appendix E.3. The g_i, which appear in these equations, are functions only of γ and w; they are listed in Appendix E.4. The equations in Equation (4.10) provide the first four derivatives in Appendix E.3. For convenience, several others have been included. In contrast to the tangential derivatives, most normal derivatives contain a σ term. This stems from the σ term in continuity.

Subsequent to the above analysis, it was realized that a simpler approach is possible, because the flow is homenergetic—that is,

$$h_o = \frac{\gamma}{\gamma-1}\frac{p}{\rho} + \frac{1}{2}u^2 + \frac{1}{2}v^2 = \text{constant}$$

This readily yields

$$\frac{1}{\rho}\frac{\partial p}{\partial n} = \frac{p}{\rho^2}\frac{\partial \rho}{\partial n} - \frac{\gamma-1}{\gamma}\left(u\frac{\partial u}{\partial n} + v\frac{\partial v}{\partial n}\right) \tag{4.11a}$$

and a similar result for the tangential pressure derivative. Thus, the energy equation and the pressure gradients in Equation (4.10) can be eliminated. After this replacement, the tangential momentum equation contains only one normal derivative term. It provides an explicit solution for $\partial u/\partial n$:

$$\frac{\partial u}{\partial n} = -\frac{u}{\gamma v}\frac{\partial u}{\partial s} + \frac{\gamma-1}{\gamma}\frac{\partial v}{\partial s} - \beta' u - \frac{p}{\rho^2 v}\frac{\partial \rho}{\partial s} \tag{4.11b}$$

Continuity and the normal momentum equation are the remaining two linear equations

$$\rho\frac{\partial v}{\partial n} + v\frac{\partial \rho}{\partial n} = -\rho\frac{\partial u}{\partial s} - u\frac{\partial \rho}{\partial s} - \rho v\beta' - \frac{\sigma\rho}{y}\left(u\sin\beta - v\cos\beta\right) \tag{4.11c}$$

$$\frac{v}{\gamma}\frac{\partial v}{\partial n} + \frac{p}{\rho^2}\frac{\partial \rho}{\partial n} = -u\frac{\partial v}{\partial s} + \frac{\gamma-1}{\gamma}u\frac{\partial u}{\partial n} + \beta' u^2 \tag{4.11d}$$

that determine $\partial v/\partial n$ and $\partial \rho/\partial n$. Analytically, Equation (4.11b) is first solved, then Equation (4.11c,d), and finally Equation (4.11a). The analytical solution has been performed by the author; it checks the one in Appendix E.3 by MACSYMA.

Several normal derivatives become infinite when $Z = 0$ or $w = 1$. These denominator Zs invariably appear as β'/Z, and the infinity is removed by setting β' equal to zero. In the $w \to 1$ limit, the shock thus becomes a Mach wave with zero longitudinal curvature. Another limit is the hypersonic one. If the shock wave is normal, or nearly normal, to the upstream flow, the rightmost terms in Appendix E.4 dominate, and the X, Y, and Z factors simplify in an obvious manner. Another hypersonic limit is for slender bodies, when w is of order unity, and the g_i, X, Y, and Z factors do not simplify. Nevertheless, the equations do simplify because of the presence of m, which approaches infinity.

5 Derivative Applications

5.1 NORMAL DERIVATIVES WHEN THE SHOCK IS NORMAL TO THE UPSTREAM VELOCITY

As evident from Appendix E, the normal derivatives are more involved than the tangential ones. There is a major simplification of the normal derivates, however, when the shock is normal to the upstream velocity, as occurs at the nose of a detached shock.

Although the shock is normal to the freestream, it may be convex or concave. Both configurations are discussed in Section 5.5, while Section 9.4 briefly discusses a shock with a saddle point. (A saddle point may occur if the body that generates a detached shock has two "noses" closely spaced. Think of two adjacent knuckles on a fist.) For the subsequent normal shock analysis, it is nevertheless convenient to first introduce some curvature concepts.

A two-dimensional parabolic or hyperbolic shock becomes a paraboloid or hyperboloid shock when axisymmetric. A convex hyperbola and hyperboloid shock are extensively discussed in Section 6.4. In general, much of the discussion has been oriented toward a detached, convex bow shock. In Section 5.5, however, a shock that is concave to a uniform upstream flow is considered.

The shock curvatures are defined as

$$\kappa_s = -\beta', \quad \kappa_t = \frac{\sigma \cos \beta}{y} \quad (5.1a,b)$$

where κ_t (t for transverse) is in a plane normal to the flow plane and the shock. It is zero when the shock is two-dimensional. The two curvatures are positive for an axisymmetric convex shock and negative when the shock is axisymmetric and concave. The concave result for κ_t stems from $\cos \beta$ being negative (β is in the second quadrant) for a concave shock. It is useful that the curvatures have the same sign, because an axisymmetric shock, where it is normal to the freestream velocity, has $\kappa_s = \kappa_t$.

When an axisymmetric shock is a normal shock, the $\cos \beta / y$ ratio is indeterminate, but it is evaluated by L'Hospital's rule as

$$lim_{s \to 0} \frac{\cos \beta}{y} = \frac{\dfrac{d \cos \beta}{ds}}{\dfrac{dy}{ds}} = -\frac{\beta' \sin \beta}{1} = -\beta' = \frac{1}{R_s} \quad (5.2)$$

where R_s is the radius of curvature of the nose of the shock. Remember, for a convex or concave axisymmetric shock, the two radii of curvatures are equal. We thus have

$$\beta = 90°, \quad \theta = 0°, \quad \beta' = -\frac{1}{R_s}, \quad \frac{\cos\beta}{y} = \frac{1}{R_s} \tag{5.3}$$

where R_s is positive for a convex shock and negative for a concave shock. The two nonzero tangential derivatives in Appendix E.2 become

$$\left(\frac{\partial u}{\partial s}\right)_2 = -\beta', \quad \left(\frac{\partial \theta}{\partial s}\right)_2 = \frac{Z}{X}(-\beta')$$

Because u_2 and θ_2 are zero for a normal shock, we see that their values positively increase from zero for a convex shock but negatively increase for a concave shock. The latter case is understood by noting that $\bar{V}_2$ slopes toward the symmetry axis.

As derived in Problem 10, the simplified results (with $w = m$) are

$$mg_3 + g_4 = -X^2 Z \tag{5.4a}$$

$$mg_5 + g_6 = -\frac{4}{\gamma+1} XYZ \tag{5.4b}$$

$$(\gamma + 1)g_2 - 2m(1 + 3m) = \frac{4}{\gamma+1} YZ \tag{5.4c}$$

$$\frac{1}{V_1}\left(\frac{\partial v}{\partial n}\right)_2 = -\left(\frac{2}{\gamma+1}\right)^2 \frac{Y}{m} \frac{1+\sigma}{R_s} \tag{5.5a}$$

$$\frac{1}{p_1}\left(\frac{\partial p}{\partial n}\right)_2 = \frac{4\gamma}{(\gamma+1)^2} Y \frac{1+\sigma}{R_s} \tag{5.5b}$$

$$\frac{1}{\rho_1}\left(\frac{\partial \rho}{\partial n}\right)_2 = \left(\frac{m}{X}\right)\frac{1+\sigma}{R_s} \tag{5.5c}$$

$$\frac{1}{T_1}\left(\frac{\partial T}{\partial n}\right)_2 = \frac{8(\gamma-1)}{(\gamma+1)^3} \frac{XY}{m} \frac{1+\sigma}{R_s} \tag{5.5d}$$

where T is temperature, and $(\partial u/\partial n)_2$ and $(\partial p_o/\partial n)_2$ are zero. The above derivatives only depend on γ, M_1^2, and $(1 + \sigma)/R_s$. Because of the R_s sign convention, the flow is thus compressive (expansive) for a convex (concave) shock. Their magnitude varies inversely with R_s and doubles for an axisymmetric shock. The derivatives are zero for a normal shock without curvature, as expected. The v derivative can be used to obtain an estimate of the shock stand-off distance (Problem 20) for a detached convex shock, which is provided in the "Generic Shock Shape" section of Chapter 6. The top three g_i equations are readily checked using prescribed γ and $w(=m)$ values with the equations in Appendix E.4.

By way of contrast, Problems 11 and 12 deal with two unsteady, normal shock flows where the above analysis does not apply.

Lin and Rubinov (1948), using curved shock theory, appear to demonstrate that a normal shock, at its foot, cannot be attached to a concave (convex) wall, unless M_1 exceeds (is below) a critical value:

$$M_c = \left\{ \frac{1}{2} \left[\gamma + 1 + \left(\gamma^2 + 2\gamma + 5 \right)^2 \right] \right\}^{1/2}$$

which is 1.662 when $\gamma = 1.4$. This assertion is now examined under a steady flow assumption, and the shock may be three-dimensional but with a uniform freestream. With a uniform freestream, the upstream streamline curvature, $\partial \theta / \partial \tilde{s}$, is obviously zero, where $\tilde{s}$ is distance along a streamline. This curvature is proportional to the streamline's normal pressure gradient, $\partial p / \partial \tilde{n}$. (In an unsteady flow, there is an additional acceleration term.) Hence, downstream of a normal shock, using the normal momentum equation, one can show that the streamline's curvature is also zero. When a normal shock is attached to a wall, the wall must have zero longitudinal curvature at the attachment point. Thus, an attached, normal shock on a curved wall is unstable, regardless of the value of the upstream supersonic Mach number. This instability implication of Lin and Rubinov's analysis tends to be borne out by a number of photographs in Van Dyke's (1982) album. The critical Mach number assertion, however, is not valid.

Photographs in the album show shocks that are normal to a longitudinally flat surface, such as a projectile's cylindrical surface. There is one photograph (number 250) that shows a normal shock over a curved surface. The foot of the shock, however, is actually on the top of a thickened, or separated, turbulent boundary layer whose upper surface is flat.

5.2   INTRINSIC COORDINATE DERIVATIVES

We start with a solid-body rotation of the x,y coordinates, as shown in Figure 5.1. This rotation is readily given by

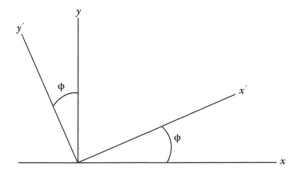

FIGURE 5.1   Solid-body rotation.

$$x' = x \cos\varphi + y \sin\varphi \qquad (5.6a)$$

$$y' = -x \sin\varphi + y \cos\varphi \qquad (5.6b)$$

or by its inversion

$$x = x' \cos\varphi - y' \sin\varphi \qquad (5.7a)$$

$$y = x' \sin\varphi + y' \cos\varphi \qquad (5.7b)$$

Intrinsic coordinates (see Figure 5.2) are introduced, where $\tilde{s}$ is along a streamline in the positive velocity direction, and $\tilde{n}$ is normal to the streamline in the osculating plane (i.e., the plane that contains $\vec{V}$ and $\vec{V} \pm d\vec{V}$ for a fixed streamline). (For a two-dimensional or axisymmetric shock, the flow and osculating planes coincide.) The $\tilde{n}$ coordinate, by its conventional definition, is positive (as shown in Figure 5.2) when in the direction of the radius of curvature vector of the streamline. For an orthonormal, right-handed, intrinsic coordinate system, $\tilde{n}, \tilde{t}, \tilde{b}$, the coordinate $\tilde{b}$ is normal to the osculating plane and is positive in the direction normal to, and into, the plane of the Figure 5.2 page. Intrinsic coordinates can be used at any point in the flow field; we apply them only to state 2.

Although the s,n and $\tilde{s}, \tilde{n}$ coordinates are curvilinear, a local transformation is utilized in which coordinates are straight. Our objective is to obtain various partial derivatives of the two coordinate systems in terms of β and θ. For this, Equations (5.7a,b) are used to rotate the n,s coordinates into the $\tilde{s}, \tilde{n}$ coordinates, respectively. To do this, use the replacement

$$x \to n, \quad y \to s, \quad x' \to \tilde{s}, \quad y' \to \tilde{n}, \quad \varphi \to 90 - (\beta - \theta)$$

where x is along $\vec{V}_1$. This yields

$$n = \tilde{s} \sin(\beta - \theta) - \tilde{n} \cos(\beta - \theta) \qquad (5.8a)$$

$$s = \tilde{s} \cos(\beta - \theta) + \tilde{n} \sin(\beta - \theta) \qquad (5.8b)$$

where φ is the angle between x and x' (i.e., between $\hat{n}$ and $\vec{V}_2$). In what follows, the state 2 subscript is not shown, except for purposes of clarity. Also note that the fixed coordinate is usually not indicated:

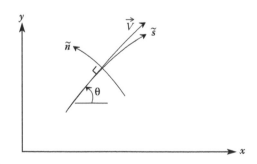

FIGURE 5.2   Intrinsic coordinates in the osculating plane, $\tilde{b}$ is into the page.

$$\frac{\partial}{\partial s} = \left(\frac{\partial}{\partial s}\right)_n, \quad \frac{\partial}{\partial n} = \left(\frac{\partial}{\partial n}\right)_s, \quad \frac{\partial}{\partial \tilde{s}} = \left(\frac{\partial}{\partial \tilde{s}}\right)_{\tilde{n}}, \quad \frac{\partial}{\partial \tilde{n}} = \left(\frac{\partial}{\partial \tilde{n}}\right)_{\tilde{s}}$$

and that the rotation angle φ (= $90 + \theta - \beta$) is also fixed. Hence, we obtain from the equations in Equation (5.8)

$$\frac{\partial s}{\partial \tilde{n}} = \sin(\beta - \theta), \quad \frac{\partial s}{\partial \tilde{s}} = \cos(\beta - \theta), \quad \frac{\partial n}{\partial \tilde{n}} = -\cos(\beta - \theta), \quad \frac{\partial n}{\partial \tilde{s}} = \sin(\beta - \theta) \quad (5.9)$$

The inverse derivatives, such as $(\partial \tilde{n} / \partial s)_n$, can be obtained from the equations in Equation (5.6). The intrinsic coordinate derivatives, $(\partial(\ )/\partial \tilde{s})_{\tilde{n}}$ and $(\partial(\ )/\partial \tilde{n})_{\tilde{s}}$, are constructed from the s and n derivatives listed in Appendix E using the chain rule

$$\frac{\partial}{\partial \tilde{s}_{\tilde{n}}} = \frac{\partial s}{\partial \tilde{s}_{\tilde{n}}} \frac{\partial}{\partial s_n} + \frac{\partial n}{\partial \tilde{s}_{\tilde{n}}} \frac{\partial}{\partial n_s}, \quad \frac{\partial}{\partial \tilde{n}_{\tilde{s}}} = \frac{\partial s}{\partial \tilde{s}_{\tilde{n}}} \frac{\partial}{\partial s_n} + \frac{\partial n}{\partial \tilde{n}_{\tilde{s}}} \frac{\partial}{\partial n_s}$$

With the equations in Equation (5.9) and Appendix E.1, this becomes

$$\left(\frac{\partial}{\partial \tilde{s}}\right)_2 = \cos(\beta - \theta)\frac{\partial}{\partial s} + \sin(\beta - \theta)\frac{\partial}{\partial n} = \frac{1}{B^{1/2}}\left(A\frac{\partial}{\partial s} + \frac{\partial}{\partial n}\right) \quad (5.10a)$$

$$\left(\frac{\partial}{\partial \tilde{n}}\right)_2 = \sin(\beta - \theta)\frac{\partial}{\partial s} - \cos(\beta - \theta)\frac{\partial}{\partial n} = \frac{1}{B^{1/2}}\left(\frac{\partial}{\partial s} - A\frac{\partial}{\partial n}\right) \quad (5.10b)$$

Downstream of a steady shock, the stagnation pressure should be constant along a streamline. As a check on the theory, the streamline derivative $(\partial p_o/\partial \tilde{s})_2$ is readily shown to be zero. Problem 8 is another independent check on the theory. The inverse of the above equations

$$\left(\frac{\partial}{\partial s}\right)_2 = \frac{1}{B^{1/2}}\left(A\frac{\partial}{\partial \tilde{s}} + \frac{\partial}{\partial \tilde{n}}\right) \quad (5.11a)$$

$$\left(\frac{\partial}{\partial n}\right)_2 = \frac{1}{B^{1/2}}\left(\frac{\partial}{\partial \tilde{s}} - A\frac{\partial}{\partial \tilde{n}}\right) \quad (5.11b)$$

is required later. Equations (5.10) and (5.11) are identical except for an interchange of s,n with $\tilde{s},\tilde{n}$.

5.3 DERIVATIVES ALONG CHARACTERISTICS

To further illustrate the theory, the differential operators along characteristics, or Mach lines, are developed (see Figures 3.1 and 5.3). Streamlines are denoted as ζ_o and have an angle θ relative to the x_1-axis, while the left-running (ζ_+) Mach lines have an angle $\mu + \theta$ and the right-running (ζ_-) lines have a positive angle $\mu - \theta$,

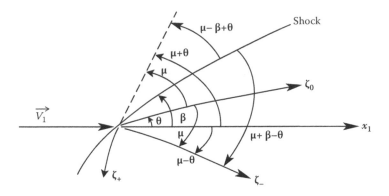

FIGURE 5.3 Angles for ζ_o and $\zeta_\pm$ relative to x_1 and the shock.

both with respect to the x_1-axis. The left- and right-running characteristic terminology stems from an observer facing in the downstream direction. A right-running characteristic, ζ_-, has the direction of an outstretched right arm. The same applies to the left-running characteristic, ζ_+, being aligned with an outstretched left arm. Because of the shock, for analytical convenience, the direction of ζ_+ is reversed in the figure. Streamlines are also denoted as $\tilde{s}$. Their derivative is provided by Equation (5.10a). Essential Mach lines have angles $\pm\mu$ with respect to a streamline. For convenience, angles are positive in the direction of the arrows in the figure. The variables ζ_o and $\zeta_\pm$ represent arc lengths in their respective directions.

A small disturbance, starting at a point downstream of the shock, can propagate in the downstream direction along the left-running characteristics in the $-\zeta_+$ direction (i.e., toward the shock wave, see Figure 5.3). The interaction between the incident wave and the shock wave has several effects on the flow. The disturbance alters the shock's slope thereby generating shock-produced vorticity. This vorticity is transported downstream along streamlines and is referred to as a *vortical layer*. The disturbance also reflects from the shock along right-running (ζ_-) characteristics, which is the topic of the next section.

For the Mach line directions, the sines and cosines of $\mu + \beta - \theta$ and $\mu - \beta + \theta$ are required. These are the angles that the ζ_- and ζ_+ characteristics have with respect to the shock (see Figure 5.3). As an example, one of the sines is evaluated:

$$\sin(\mu+\beta-\theta)=\sin(\beta-\theta)\cos\mu+\cos(\beta-\theta)\sin\mu = \frac{1}{M_2 B^{1/2}}\left[\left(M_2^2-1\right)^{1/2}+A\right]$$

$$=\left(\frac{Y}{X}\right)^{1/2}\frac{A+\left(M_2^2-1\right)^{1/2}}{B}$$

where it is not convenient to eliminate $\left(M_2^2-1\right)^{1/2}$. Note that M_2 must equal or exceed unity for a real-valued result. The analysis, therefore, does not apply to any part of the shock where the state 2 flow is subsonic. In a similar manner, we obtain

$$\cos(\mu + \beta - \theta) = \left(\frac{Y}{X}\right)^{1/2} \frac{A\left(M_2^2 - 1\right)^{1/2} - 1}{B}$$

$$\sin(\mu - \beta + \theta) = \left(\frac{Y}{X}\right)^{1/2} \frac{A - \left(M_2^2 - 1\right)^{1/2}}{B}$$

$$\cos(\mu - \beta + \theta) = \left(\frac{Y}{X}\right)^{1/2} \frac{A\left(M^2 - 1\right)^{1/2} + 1}{B}$$

As with the equations in Equation (5.9), the right-running characteristic direction utilizes

$$\left(\frac{\partial s}{\partial \zeta_-}\right)_2 = \cos(\mu + \beta - \theta), \quad \left(\frac{\partial n}{\partial \zeta_-}\right)_2 = \sin(\mu + \beta - \theta)$$

with the result

$$\left(\frac{\partial}{\partial \zeta_-}\right)_2 = \left(\frac{\partial s}{\partial \zeta_-}\frac{\partial}{\partial s} + \frac{\partial n}{\partial \zeta_-}\frac{\partial}{\partial n}\right)_2$$

$$= \left(\frac{Y}{X}\right)^{1/2} \frac{1}{B}\left\{\left[A\left(M_2^2 - 1\right)^{1/2} - 1\right]\left(\frac{\partial}{\partial s}\right)_2 + \left[\left(M_2^2 - 1\right)^{1/2} + A\right]\left(\frac{\partial}{\partial n}\right)_2\right\}$$

(5.12a)

The left-running characteristic direction is given by

$$\left(\frac{\partial}{\partial \zeta_+}\right)_2 = \left(\frac{\partial s}{\partial \zeta_+}\frac{\partial}{\partial s} + \frac{\partial n}{\partial \zeta_+}\frac{\partial}{\partial n}\right)_2 = -\cos(\mu - \beta + \theta)\left(\frac{\partial}{\partial s}\right)_2 + \sin(\mu - \beta + \theta)\left(\frac{\partial}{\partial n}\right)_2$$

$$= -\left(\frac{Y}{X}\right)^{1/2} \frac{1}{B}\left\{\left[A\left(M_2^2 - 1\right)^{1/2} + 1\right]\left(\frac{\partial}{\partial s}\right)_2 + \left[\left(M_2^2 - 1\right)^{1/2} - A\right]\left(\frac{\partial}{\partial n}\right)_2\right\}$$

(5.12b)

Equation (5.12) provides exact, explicit relations for the two derivatives in the flow plane. The comparable streamline derivative is provided by Equation (5.10a), which is appreciably simpler than the above relations.

When state 2 is sonic, the above equations reduce to

$$\left(\frac{\partial}{\partial \zeta_-}\right)_2 = \left(\frac{\partial}{\partial \zeta_+}\right)_2 = \frac{1}{B^{1/2}}\left(-\frac{\partial}{\partial s} + A\frac{\partial}{\partial n}\right)$$

which is the negative of $\partial(\ )/\partial \tilde{n}$ given by Equation (5.10b). By definition, the derivatives are positive along their respective characteristics, but here point in opposite directions. For a sonic point on a convex shock, the right-running characteristic points into a subsonic flow and has zero length. The left-running characteristic points upstream, into the shock, and also has zero length. A zero length is equivalent to zero strength.

5.4   WAVE REFLECTION FROM A SHOCK WAVE

As mentioned, left-running Mach lines reflect, in part, from the downstream side of a convex shock as a wave consisting of right-running Mach lines. The reflected wave is an expansion wave if its Mach lines diverge from each other. If they converge, the wave is compressive. Moreover, converging Mach lines that attempt to overlap form a weak shock wave where the overlap would occur. Thus, an internal shock can form in a supersonic flow containing converging Mach lines of the same family. In this situation, flow conditions upstream of the internal shock are nonuniform. This process results in the downstream shock system that appears in a jet emanating from an underexpanded nozzle (Emanuel 1986, Section 19.4).

If the incident wave is compressive, its interaction will strengthen the shock causing β' to be less negative. An inflection point on the shock (Wilson 1967) would occur if the compression is strong enough to cause β' to become positive. This can occur, for example, if a slightly convex wedge or spike, with an attached shock, has a concave change in shape. If the incident wave is expansive, it weakens the shock. In either case, there are two reflected waves, the one with right-running Mach lines and a vortical, streamline layer due to the induced change in the shock's curvature. The strength of both reflected waves depends on the change of shock curvature caused by the incident wave.

The family of right-running ζ characteristics is referred to as a C_- wave. The slope of these characteristics, just downstream of the shock, is $\mu - \theta$ relative to the x_1-axis, as previously mentioned. By traveling along a convex shock, in the downstream direction, the wave is seen to be compressive (expansive) if the positive angle, $\mu - \theta$, increases (decreases). (When $\mu - \theta$ increases, Figure 5.3 shows that the ζ characteristics are converging.) Thus, the C_+ wave reflects from a shock as a compression if

$$\frac{d(\mu-\theta)}{ds} > 0 \quad \text{or} \quad \frac{d(\mu-\theta)}{d\beta} < 0$$

where the second form is analytically more convenient.

It is difficult to derive an equation for $d(\mu - \theta)/d\beta$ without the assistance of the theory in Chapter 4. With this theory, the derivation is straightforward; start with Equation (4.6) to obtain

$$\frac{d\mu}{dM} = -\frac{1}{M(M^2-1)^{1/2}}$$

Consequently, write the derivative of interest as

$$\frac{d(\mu-\theta)}{d\beta} = \frac{\dfrac{d\mu}{dM}\dfrac{dM}{ds} - \dfrac{d\theta}{ds}}{\dfrac{d\beta}{ds}} = -\frac{1}{\beta'}\left[\frac{1}{2M^2(M^2-1)^{1/2}}\frac{dM^2}{ds} + \frac{d\theta}{ds}\right] \qquad (5.13a)$$

With the assistance of Appendix E, this becomes

$$\frac{d(\mu-\theta)}{d\beta} = \frac{1}{X^2B}\left[\frac{\left(\dfrac{\gamma+1}{2}\right)\left(1+\dfrac{\gamma-1}{2}m\right)(1+\gamma w^2)A}{Y\left(\dfrac{X}{Y}B-1\right)^{1/2}} - \frac{\gamma+1}{2}m(1+w)-1+2w+\gamma w^2\right]$$

$$(5.13b)$$

This is an exact result that is independent of whether or not the flow is two-dimensional or axisymmetric. It is also independent of the local shock wave curvature, $-\beta'$, because this parameter cancels. Although complicated, the right side only depends on γ, M_1, and β; hence, the influence of the incident wave is limited to its effect on the wave angle β. Moreover, the larger the magnitude of the derivative, the stronger is the reflected expansion or compression. In arriving at the above result, it is useful to note that (see Problem 6)

$$X^2B = -(1+\gamma w)Z+\left(\frac{\gamma+1}{2}\right)^2 mw = (\gamma+1)\left(1+\frac{\gamma+1}{4}m\right)w+1-2w-\gamma w^2 \quad (5.13c)$$

For a detached shock, the flow between the shock and body in the nose region is subsonic. This region is bordered by a curved sonic line that intersects the shock where its slope is β^*. The above convex shock analysis, of course, only holds when β is less than β^*. At β^*, Equation (5.13a) shows that $d(\mu-\theta)/d\beta$ is infinite, because M_2 equals unity. A relation for β^* is obtained by setting $M_2=1$ (see Appendix E.1):

$$(\gamma+1)mw^* +2+(\gamma-3)w^* -2\gamma(w^*)^2 = 0$$

This relation becomes

$$\gamma(M_1\sin\beta^*)^4 -\frac{1}{2}\left[\gamma-3+(\gamma+1)M_1^2\right](M_1\sin\beta^*)^2 -1 = 0$$

which is a quadratic equation for $(M_1\sin\beta^*)^2$, with the result

$$\sin^2\beta^* = \frac{\gamma+1}{4\gamma M_1^2}\left\{M_1^2 -\frac{3-\gamma}{\gamma+1}+\left[M_1^4 -\frac{2(3-\gamma)}{\gamma+1}M_1^2 +\frac{\gamma+9}{\gamma+1}\right]^{1/2}\right\} \qquad (5.14)$$

With $\gamma = 1.4$ and $1.59 \leq M_1 \leq \infty$, β^* is confined to the narrow $61.70°$ to $67.79°$ range, where the second value occurs when M_1 is infinite.

As is often the case, the incoming wave is an expansion, thereby weakening the shock. In a blunt body flow, the C_+ wave originates on the sonic line. In any case, detailed calculations (see Problem 15) with $\gamma = 1.4$ show that the reflected wave is expansive, for all β values, when $M_1 < 1.59$. For larger M_1 values, there is a range of β values:

$$\mu_1 \leq \beta \leq 39°, \quad M_1 = 1.59$$

$$39°-, \quad 2 \leq M_1 \leq 4$$

$$38°+, \quad M_1 = 6$$

$$38°+, \quad M_1 = 8$$

for which the reflected wave is compressive, where μ_1 is the state 1 Mach angle. The compressive β region starts at $M_1 = 1.59$, where $\mu_1(1.59) = 39.0°$. As evident, the upper limit for β, where $d(\mu - \theta)/d\beta = 0$, decreases quite slowly with M_1. On the other hand, μ_1 rapidly decreases, thereby increasing the range of relatively small β values for which the reflected wave is compressive. For instance, when $M_1 = 4$, the reflected wave is compressive when β is between $14.48°$ and $38°+$. For larger β values, the reflected wave is expansive. (This expansive wave can then interact with the boundary layer on the surface of the vehicle.) Consequently, for a freestream Mach number in excess of 1.59, both types of reflection processes are present, as sketched in Figure 5.4. Note that the compressive reflection occurs downstream, where the shock is weak. On the other hand, at the sonic point, where $d(\mu - \theta)/d\beta$ is infinite, the reflected right-running wave is expansive, but, initially, is of zero strength.

Hypersonic small disturbance theory is now briefly discussed. In this theory, we have the limit

$$M_\infty \to \infty, \quad K_\beta = M_\infty \sin\beta = O(1)$$

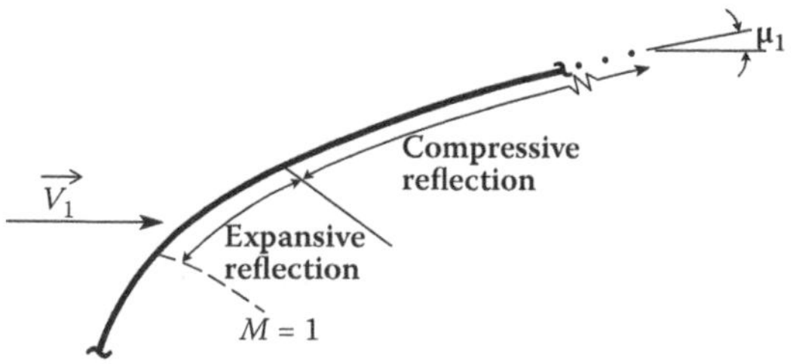

FIGURE 5.4 Expansive and compressive regions downstream of a shock when $\gamma = 1.4$ and $M_1 > 1.59$.

with

$$X = 1 + \frac{\gamma - 1}{2} K_\beta^2, \quad Y = \gamma K_\beta^2 - \frac{\gamma - 1}{2}$$

Equation (5.13b) yields, to leading order,

$$\frac{d(\mu - \theta)}{d\beta} = \frac{\gamma - 1}{\gamma + 1} \frac{1 + \gamma K_\beta^4}{K_\beta^2 X^{1/2} Y^{1/2}} - \frac{2}{\gamma + 1} \frac{1 + K_\beta^2}{K_\beta^2}$$

For instance, at a point on the shock where K_β is unity, this becomes

$$\frac{d(\mu - \theta)}{d\beta} = -\frac{2(3 - \gamma)}{\gamma + 1}$$

and the reflected wave, at this location, is compressive.

5.5 FLOWS WITH A CONICAL SHOCK WAVE

As we know, the tangential and normal derivatives are zero downstream of a straight shock that is attached to a wedge. For a cone at zero incidence with an attached conical shock, the tangential derivatives are again zero while the normal derivatives greatly simplify (see Problem 16). The shock has β' equal to zero and σ equal to unity. This flow, known as Taylor-Maccoll flow, is discussed in Section 9.6 of Emanuel (2001), where it is shown that the Euler equations of motion reduce to two coupled first-order ordinary differential equations (ODEs) whose independent variable is the angle η (see Figure 9.27 in Emanuel 2001). (In this and the next few paragraphs, the discussion describes material in Chapter 9 of the above reference.) Chapter 9 is devoted to calorically imperfect gas flows where Taylor-Maccoll flow is one example. Section 9.6, however, contains a short subsection with the perfect gas formulation. (Problems 16 and 17 also deal with Taylor-Maccoll flow of a perfect gas.)

In Taylor-Maccoll flow, the flow is irrotational, homentropic, and depends on a single angular variable. It is the axisymmetric counterpart to a Prandtl-Meyer expansion or compression. It differs from a Prandtl-Meyer flow in that it also applies when the downstream flow is subsonic. As will be shown, the flow about a conical body is compressive. Later in the discussion, an expansive Taylor-Maccoll flow is encountered.

In Figure 9.27, θ_b is the cone's half angle, the radial coordinate r is replaced with y, and the u and v velocity components are defined differently. The flow just downstream of the shock is usually supersonic but can also be subsonic. Between the shock and body the flow may be entirely supersonic, subsonic, or mixed. When mixed, there is a sonic conical surface. Disturbances propagate (and attenuate) in the upstream direction when some, or all, of the flow is subsonic. There would be a disturbance (e.g., caused by the shoulder where the base of a cone is attached to a sting support). When some of the flow is subsonic, the Taylor-Maccoll solution asymptotically holds as the cone's apex is approached.

The θ_b, β variation, for several M_1 values with $\gamma = 1.4$, is shown in Figure 9.28b, where the perfect gas solution is the solid $\delta = 0$ curves. The corresponding wedge result is shown in Figure 9.28a. In both cases, the attached, weak solution shock is to the left of the maximum of the curves. Note that the θ_b value for detachment substantially exceeds its wedge counterpart. Because β^* (given by Equation 5.14) is independent of dimensionality, the θ_b range of values, between where $M_2 = 1$ and detachment occurs, is appreciably larger than in the planar case. This feature is evident in Figure 4 of NACA 1135 (Ames Research Staff, 1953).

From Problem 16 and Equation (5.25), given later, we obtain

$$\frac{1}{p_1}\left(\frac{\partial p}{\partial n}\right)_2 = \left(\frac{2}{\gamma+1}\right)^2 \gamma Y \frac{\cos\beta}{y} \tag{5.15a}$$

$$\left(\frac{\partial M^2}{\partial n}\right)_2 = -(\gamma+1)\frac{\left(1+\frac{\gamma-1}{2}m\right)w}{XY}\frac{\cos\beta}{y} \tag{5.15b}$$

$$\left(\frac{\partial\theta}{\partial\tilde{s}}\right)_2 = \frac{2}{\gamma+1}\frac{A}{B^{3/2}}\frac{Y}{X}\frac{\cos\beta}{y} \tag{5.15c}$$

With β in the first quadrant, the right sides of Equation (5.15a,c) are positive, while that of Equation (5.15b) is negative. The flow is thus compressive even when M_2 is subsonic. The streamline angle θ gradually increases from θ_2 to θ_b. Because of the singularity at the cone's apex, when $y = 0$, $\theta = \theta_b$ on the cone's surface.

There is a second type of conical flow that is associated with what is conveniently referred to as an inverted cone. The author is grateful to S. Molder for his enlightening comments on this topic (see Molder 1967). A sketch of the configuration is shown in Figure 5.5, where the body is part of a hollow cylinder. At the surface's tip, the radius is y_t and the internal wall angle is θ_{bt} (b for body, t for tip). In the lower half of the figure, straight rays (actually conical surfaces) are sketched. Along a ray, the Taylor-Maccoll solution is constant for M, p, θ, ... The downstream-most ray has $\theta = 0°$ where the velocity is parallel to the x-coordinate. The 1, 2, and 3 designation applies to the upstream flow, the flow just downstream of the shock, and the parallel flow downstream of the $\theta = 0°$ ray.

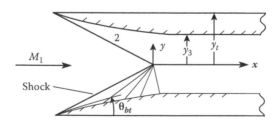

FIGURE 5.5 An inverted conical shock.

The solution in the upper region, marked with a 2, is identical to that in the lower region, except that θ_2 and $(\partial\theta/\partial\tilde{s})_2$ change sign. (The same sign change occurs with a Taylor-Maccoll flow.) In the upper region, θ_2 is negative, $(\partial\theta/\partial\tilde{s})_2$ is positive, and β is in the second quadrant. From Equation (5.15a,b), we see that $(\partial p/\partial n)_2$ is negative, due to the $\cos\beta$ factor, and $(\partial M^2/\partial n)_2$ is positive. In contrast to the earlier compressive conical flow, this flow is expansive. Because of the A factor in Equation (5.15c), the sign of $(\partial\theta/\partial\tilde{s})_2$ is determined by $\sin\beta$, not $\cos^2\beta$, and this derivative is positive in the upper region. In the upper region, the wall slope, determined by θ, gradually increases from a negative θ_{bt} value to zero.

It is not intuitive that the flow downstream of an inverted conical shock is expansive because the cross-sectional area, from region 1 to 3, goes through a contraction. For instance, with isentropic, supersonic nozzle flow there would be an area increase. The flow under discussion, however, is not isentropic or one-dimensional.

The length ratio, y_t/y_3, is determined by a combination of the conservation of mass flow rate and the Taylor-Maccoll equations. We start by noting that

$$T_{o1} = T_{o2} = T_{o3}, \quad p_{o2} = p_{o3}, \quad \rho_{o2} = \rho_{o3}$$

where the oh subscript denotes a stagnation value. For a uniform flow at states 1 and 3, the mass flow rate is written as

$$\dot{m} = \rho A V = \rho_o \left(\gamma R T_o\right)^{1/2} A \frac{M}{\left(1 + \dfrac{\gamma-1}{2} M^2\right)^{(\gamma+1)/[2(\gamma-1)]}}$$

where A is the cross-sectional area. With several gas dynamic relations and $\dot{m}_1 = \dot{m}_3$, we obtain

$$\left(\frac{y_t}{y_3}\right)^2 = \frac{\gamma+1}{2} \frac{M_{1n}^2}{1 + \dfrac{\gamma-1}{2} M_{1n}^2} \frac{M_3}{M_1} \left(\frac{1 + \dfrac{\gamma-1}{2} M_1^2}{1 + \dfrac{\gamma-1}{2} M_3^2}\right)^{(\gamma+1)/[2(\gamma-1)]} \tag{5.16}$$

Values are prescribed for γ, M_1, and β, which determine M_{1n}, M_2, θ_{bt}, p_2, ... The unknowns in Equation (5.16) are M_3 and y_t/y_3. The Taylor-Maccoll equations determine M_3.

For purposes of simplicity and clarity, there is one inverse cone aspect that has not been mentioned. The flow configuration in Figure 5.5 has a central Mach disk instead of a focal point (Ferri, 1954). The reason is that the velocity, $\bar{V}_2$, points toward the symmetry axis. Consequently, there is either a Mach disk or a second conical shock, whose apex is at the origin, that would turn the flow parallel to the x-axis. For the Mach disk case, there is a triple point (actually circular line) where the Mach disk and incident and reflected shocks meet. Triple points are the subject of Chapter 7. The upstream flow with a second conical shock somewhat resembles the flow in a Busemann diffuser (Molder, 1967). With a Busemann diffuser, the upstream shock becomes a Mach cone and the flow is compressive, not expansive. Both flows use the Taylor-Maccoll equations, but the rays are oppositely oriented such that the converging walls have curvatures of the opposite sign. The author is unaware if this double conical shock flow has been experimentally observed.

5.6   SPECIAL STATES

A number of special state 2 points are discussed. The first four hold generally; the flow may be unsteady and three-dimensional. The last two are the Crocco and Thomas points. They require the use of normal derivatives and are thus more constrained. The presence of these various points, or states, generally requires a shock with, at least, a finite longitudinal curvature. For instance, a planar or conical shock does not possess Crocco or Thomas points.

The first state is where the shock is normal to the freestream and is the subject of Section 5.1. The second state is where $M_2 = 1$ (i.e., the sonic state, whose β^* value is given by Equation 5.14).

For a perfect gas, the β,θ angles, in the flow plane, are related by Equation (2.28), where θ has a maximum value when β has its detachment value, β_d, given later by Equation (7.10). The corresponding θ_d value can be obtained by substituting β_d into Equation (2.28). At detachment, the derivative, $(d\theta/d\beta)_d$, is zero. From Equation (8.18), with $M_{1s} = 0$, $(d\theta/d\beta)_d = 0$ yields

$$\frac{\gamma+1}{2} m \left(1+w\right) + 1 - 2w - \gamma w^2 = 0$$

which simplifies to Equation (7.10). Recall that β,θ are measured relative to the upstream velocity $\vec{V}_1$. When this velocity is not uniform, the detachment condition still occurs when θ has a maximum value, $(\partial\theta/\partial s) = 0$, as discussed below in Equation (8.18).

The fourth special state is where the vorticity, ω_2, has an extremum value. With a uniform freestream, this location is implicitly given by Equation (6.33). When the shock is convex, this extremum value is a maximum.

The Crocco point is defined by

$$\left(\frac{\partial\theta}{\partial\tilde{s}}\right)_2 = 0 \tag{5.17}$$

where the equation for $(\partial\theta/\partial\tilde{s})_2$ is given in Problem 14, or by Equation (5.25), for a two-dimensional or axisymmetric shock. As discussed in the problem, the Crocco point, for a two-dimensional shock, is given by a cubic equation in $\sin^2\beta_{cp}$. The derivative, $\partial\theta/\partial\tilde{s}$, pertains to the curvature of a streamline in the osculating plane. (For a two-dimensional or axisymmetric shock, the flow and osculating planes coincide.) When $(\partial\theta/\partial\tilde{s})_2$ is positive, the streamline, just downstream of the shock, curves upward; when negative, it curves downward. In the two-dimensional case,

Problem 14 demonstrates the closeness of the β^*, β_{cp}, and β_d values. Since β_{cp} is between β^* and β_d, the Crocco point occurs for a weak solution shock but with M_2 slightly supersonic.

Problem 14 is for selected γ and M_1 values; it is not a general demonstration that there is a real, unique Crocco point when the shock is two-dimensional. This demonstration might be done using symbolic manipulation software to show, when the $\sin^2\beta_{cp}$ cubic is written in the form

$$x^3 + ax + b = 0$$

that its discriminant, $(b/2)^2 + (a/3)^3$, is positive.

For an axisymmetric shock, the Crocco point is given by (see Problem 14)

$$\beta'_{cp} = -\frac{C_2}{G_{cp}} = -\frac{XYZ\cos\beta_{cp}}{yG_{cp}} \tag{5.18}$$

If the shock is convex,

$$\frac{XYZ\cos\beta_{cp}}{y} > 0, \quad \beta' < 0$$

and G_{cp} must be positive for a Crocco point to exist. However, G_{cp} can be positive or negative. For instance, a strong solution shock yields a G_{cp} that is negative, and the shock does not have a Crocco point. Hence, whenever an axisymmetric shock is a strong solution shock, the state 2 streamline is straight or curves upward.

The Thomas point is defined by

$$\left(\frac{\partial p}{\partial \tilde{s}}\right)_2 = 0 \tag{5.19}$$

where on one side of the point the flow is compressive while on the other side it is expansive. The name was suggested by Molder to commemorate the contributions of Thomas, such as his (1947, 1948) papers. With the assistance of Equation (5.10a) and Appendix E, we obtain

$$\frac{1}{p_1}\left(\frac{\partial p}{\partial \tilde{s}}\right)_2 = \frac{1}{B^{1/2}}\left[A\frac{1}{p_1}\left(\frac{\partial p}{\partial s}\right)_2 + \frac{1}{p_1}\left(\frac{\partial p}{\partial n}\right)_2\right]$$

$$= \frac{\gamma}{XB^{1/2}}\left\{\left[2(m\sin\beta\cos\beta)^2 + \frac{1}{\gamma+1}\frac{mg_5 + g_6}{Z}\right]\beta'\right.$$

$$\left. + \left(\frac{2}{\gamma+1}\right)^2 XY\frac{\sigma\cos\beta}{y}\right\} \tag{5.20}$$

As with a Crocco point, the Thomas point derivative is proportional to the $-\beta'$ and $\sigma\cos\beta/y$ curvatures.

For simplicity, the following Thomas point remarks are limited to a smooth, detached, convex shock. We know that $\partial p/\partial\tilde{s}$ is positive (i.e., compressive) downstream of a normal shock. As shown by Equation (5.5b), the positive value of $(\partial p/\partial n)$ $[=\partial p/\partial\tilde{s}]$ is twice as large for an axisymmetric normal shock as compared to a two-dimensional one. One can show that the shock far downstream, for either dimensionality, becomes a Mach wave, where

$$w = 1, \quad \beta = \mu_1, \quad \frac{\beta'}{Z} \to 0, \quad y \to \infty$$

In this circumstance, one can show that $\partial p/\partial\tilde{s}$ goes to zero. At these two extremes, $\partial p/\partial\tilde{s}$ is positive and zero. The occurrence of a Thomas point, however, is more problematic for an axisymmetric shock because of the factor of two.

The existence of a two-dimensional Thomas point is now examined. From Equation (5.20), and excluding the trivial case of $\beta' = 0$, it requires

$$mg_5 + g_6 = -2\,(\gamma+1)\,Z\,(m\,\sin\beta\,\cos\beta)^2$$

or

$$mg_5 + g_6 + 2\,(\gamma+1)\,w\,(w-1)\,(m-w) = 0$$

where $m \geq w \geq 1$. This relation can be written as

$$m_{tp} = \frac{-g_6 + 2\,(\gamma+1)\,w^2\,(w-1)}{g_5 + 2\,(\gamma+1)\,w(w-1)} \tag{5.21}$$

where tp stands for Thomas point. When $\gamma = 1$ and 1.4, the $m_{tp} \geq w > 1$ condition can be shown to hold, thereby demonstrating a unique Thomas point for these γ values. Moreover, since $(m_{tp}/w) = (\sin^2\beta_{tp})^{-1}$ only slightly exceeds unity, the Thomas point occurs for a strong solution shock. This is further confirmed by Problem 25. In contrast to the above generality, the possible occurrence of a Thomas point, for an axisymmetric shock, depends on a specific shock shape because of the presence of y in Equation (5.20), now written as

$$\frac{1}{p_1}\frac{\partial p}{\partial\tilde{s}} = \frac{\gamma}{XB^{1/2}}\left[G\,\beta' + \left(\frac{2}{\gamma+1}\right)^2 XY\frac{\cos\beta}{y}\right]$$

where G represents the coefficient of β' in Equation (5.20). The various factors on the right side are nonnegative, except for β' and G. Since β' is negative, G must be positive if $\partial p/\partial\tilde{s}$ has any possibility of being zero. In view of the earlier discussion, G is most likely sufficiently positive, for a Thomas point, when the shock is weaker relative to the Thomas point condition for its two-dimensional counterpart. This is illustrated by Problem 24, which evaluates the location of the Thomas point for the generic shock shape given later by Equation (6.21). (See the problem statement for specific conditions.) When the shock is two-dimensional, the strong solution Thomas point is well removed from the detachment point. In the axisymmetric case, the point occurs where the shock is also a strong solution but now is closer to the detachment point. This topic is further discussed in Section 6.6.

5.7 θ DERIVATIVES

In contrast to other variables, such as the pressure, the derivatives of θ require special treatment. This is because θ is defined by Equation (2.28) and only the $(\partial\theta/s)_2$ derivative is obtained from this equation. The other derivatives require the use of the Euler equations. In Equation (2.28), θ is the included angle between $\vec{V}_1$ and $\vec{V}_2$, whereas in the Euler equations it is the angle of $\vec{V}$ relative to an x-coordinate (see Figure 5.2). At the shock, this difference is accounted for by requiring that the x-coordinate be parallel to $\vec{V}_1$.

The four derivatives

$$
\frac{\partial\theta}{\partial s} = \left[\left(\frac{\partial\theta}{\partial s}\right)_n\right]_2, \quad
\frac{\partial\theta}{\partial\tilde{s}} = \left[\left(\frac{\partial\theta}{\partial\tilde{s}}\right)_{\tilde{n}}\right]_2, \quad
\frac{\partial\theta}{\partial\tilde{n}} = \left[\left(\frac{\partial\theta}{\partial\tilde{n}}\right)_{\tilde{s}}\right]_2, \quad
\frac{\partial\theta}{\partial n} = \left[\left(\frac{\partial\theta}{\partial n}\right)_s\right]_2
$$

are evaluated in the order listed. The equation for $(\partial\theta/\partial s)$ is given in Appendix E.2, where X^2B is provided by Equation (5.13c). As already noted, the numerator on the right side is zero at the detachment state. The equation for $(\partial\theta/\partial s)$ is defined in the flow plane but holds for an unsteady shock in a three-dimensional flow.

For $(\partial\theta/\partial\tilde{s})$, the momentum equation, transverse to a streamline,

$$
\rho V^2 \frac{\partial\theta}{\partial\tilde{s}} + \frac{\partial p}{\partial\tilde{n}} = 0 \tag{5.22}
$$

is utilized. With Equation (5.10b) and Appendix E, we have

$$
\frac{1}{p_1}\frac{\partial p}{\partial\tilde{n}} = \frac{1}{B^{1/2}}\left(\frac{1}{p_1}\frac{\partial p}{\partial s} - A\frac{1}{p_1}\frac{\partial p}{\partial n}\right)
$$

which becomes

$$
\frac{1}{p_1}\frac{\partial p}{\partial\tilde{n}} = -\frac{2\gamma}{\gamma+1}\frac{m\sin\beta\cos\beta}{X^2ZB^{1/2}}\left(G_{cp}\,\beta' + XYZ\frac{\sigma\cos\beta}{y}\right) \tag{5.23a}
$$

where

$$
G_{cp} = \frac{\gamma+1}{4}\left(mg_5 + g_6\right) - 2X^2Z \tag{5.23b}
$$

With V_2/V_1 given in Appendix E.1 and

$$
\frac{p_1}{p_2 V_2^2} = \frac{\gamma+1}{2\gamma}\frac{1}{XB} \tag{5.24}
$$

the desired derivative is

$$\frac{\partial \theta}{\partial \tilde{s}} = \frac{2\gamma}{\gamma+1} \frac{A}{X^2 ZB^{3/2}} \left(G_{cp} \, \beta' + XYZ \frac{\sigma \cos\beta}{y} \right) \tag{5.25}$$

which is also given in Problem 14.

For $(\partial\theta/\partial\tilde{n})$, start with continuity for a two-dimensional or axisymmetric flow

$$\frac{\partial \left(\rho V y^\sigma \right)}{\partial \tilde{s}} + \rho V y^\sigma \frac{\partial \theta}{\partial \tilde{n}} = 0$$

which is rewritten as

$$\frac{\partial \theta}{\partial \tilde{n}} = -\frac{\sigma}{y} \frac{\partial y}{\partial \tilde{s}} - \frac{1}{V} \frac{\partial V}{\partial \tilde{s}} - \frac{1}{\rho} \frac{\partial \rho}{\partial \tilde{s}} \tag{5.26}$$

The connection between y, when measured from a point on the shock, and the $\tilde{s}, \tilde{n}$ coordinates uses Equation (5.7b) with

$$x' \to \tilde{s}, \quad y' \to \tilde{n}, \quad x \to x, \quad y \to y, \quad \varphi \to \theta$$

$$y = \tilde{s} \, \sin\theta + \tilde{n} \, \cos\theta$$

to obtain the result

$$\frac{\partial y}{\partial \tilde{s}} = \sin\theta$$

Continuity therefore contains a term, $\sigma\sin\theta/y$, that should not be confused with the curvature $\sigma\cos\beta/y$ term. However, with Appendix E.1, we have

$$\frac{\sigma \sin\theta}{y} = \frac{Z}{XB^{1/2}} \frac{\sigma \cos\beta}{y} \tag{5.27}$$

From Appendix E and Equation (5.10b), we have

$$\frac{1}{\rho_1} \frac{\partial \rho}{\partial \tilde{s}} = \frac{1}{B^{1/2}} \left(A \frac{1}{\rho_1} \frac{\partial \rho}{\partial s} + \frac{1}{\rho_1} \frac{\partial \rho}{\partial n} \right)$$

$$= \frac{w}{X^3 ZB^{1/2}} \left\{ \left[(mg_3 + g_4) + \frac{(\gamma+1)^2}{2} (m-w)Z \right] \beta' + X^2 Z \frac{\sigma \cos\beta}{y} \right\} \tag{5.28}$$

What actually appears in Equation (5.26) is

$$\frac{1}{\rho_2}\frac{\partial\rho}{\partial\tilde{s}} = \frac{2}{\gamma+1}\frac{1}{X^2ZB^{1/2}}\left\{\left[(mg_3+g_4)+\frac{(\gamma+1)^2}{2}(m-w)Z\right]\beta' + X^2Z\frac{\sigma\cos\beta}{y}\right\}$$

(5.29)

For the V term in Equation (5.26), utilize

$$\left(\frac{\partial V}{\partial\tilde{s}}\right)_2 = \frac{\partial\left(u_2^2+v_2^2\right)^{1/2}}{\partial\tilde{s}} = \frac{1}{V_2}\left[u_2\left(\frac{\partial u}{\partial\tilde{s}}\right)_2 + v_2\left(\frac{\partial v}{\partial\tilde{s}}\right)_2\right]$$

Again, with Appendix E and Equation (5.10b), we obtain

$$\frac{1}{V_2}\frac{\partial V}{\partial\tilde{s}} = \frac{\gamma+1}{2}\frac{1}{XZB^{3/2}}\left\{\left[g_2 - \frac{2}{\gamma+1}m\,(1+3w)\left(1+\frac{\gamma+1}{2}\frac{Z\cos^2\beta}{X}\right)\right]\beta'\right.$$

$$\left. -\left(\frac{2}{\gamma+1}\right)^2 YZ\frac{\sigma\cos\beta}{y}\right\}$$

(5.30)

The final result is obtained by combining the above with Equation (5.26):

$$\frac{\partial\theta}{\partial\tilde{n}} = -\frac{1}{X^2ZB^{3/2}}\left(\frac{\gamma+1}{2}XG_3 + \frac{2}{\gamma+1}BG_4\right)\beta'$$

$$-\frac{2}{\gamma+1}\frac{1}{XB^{1/2}}\left(\gamma w - \frac{\gamma-1}{2} - \frac{Y}{B}\right)\frac{\sigma\cos\beta}{y}$$

(5.31)

where

$$G_3 = g_2 - \frac{2}{\gamma+1}m(1+3w)\left(1+\frac{\gamma+1}{2}\frac{Z\cos^2\beta}{X}\right)$$

(5.32a)

$$G_4 = mg_3 + g_4 + \frac{(\gamma+1)^2}{2}(m-w)Z$$

(5.32b)

The final derivative, $\partial\theta/\partial n$, utilizes Equation (5.11b):

$$\frac{\partial\theta}{\partial n} = \frac{1}{B^{1/2}}\left(\frac{\partial\theta}{\partial\tilde{s}} - A\frac{\partial\theta}{\partial\tilde{n}}\right)$$

With the aid of Equations (5.25), (5.27), and (5.31), this results in

$$\frac{\partial\theta}{\partial n} = \frac{A}{XB}\left[\frac{G_5}{XZB}\beta' + \frac{2}{\gamma+1}\left(\gamma w - \frac{\gamma-1}{2}\right)\frac{\sigma\cos\beta}{y}\right]$$

(5.33)

where

$$G_5 = \frac{2}{\gamma+1} G_{cp} + \frac{\gamma+1}{2} XG_3 + \frac{2}{\gamma+1} BG_4 \qquad (5.34)$$

The four derivatives are given by Appendix E.2 and Equations (5.25), (5.31), and (5.33). They have dimensions of radians per unit length. The derivative, $\partial\theta/\partial s$, is proportional to β', the others are proportional to β' and $\sigma\cos\beta/y$.

The derivatives are evaluated for the elliptic paraboloid shock fully discussed in Section 9.3. The two-dimensional parabolic and axisymmetric hyperbolic configuration is used, with the parameters

$$\gamma=1.4, \quad M_1=3, \quad w=4, \quad r=r_2=r_3=2, \quad \sigma=0,1$$

and where f in Equation (3.2) is

$$f = (2\ rx_1)^{1/2}$$

This results in

$$\beta=41.81°, \quad \theta=23.27°, \quad M_2=1.816$$

$$x_1 = \begin{cases} 1.25, & \sigma = 0 \\ 1.25, & \sigma = 1 \end{cases}, \quad x_2 = \begin{cases} 2.236 \\ 1.581 \end{cases}, \quad x_3 = \begin{cases} 0 \\ 1.581 \end{cases}, \quad y = 2.236$$

$$\beta' = -0.1481, \quad \frac{\sigma\cos\beta}{y} = \begin{cases} 0 \\ 0.3333 \end{cases}, \quad \frac{\sigma\sin\theta}{y} = \begin{cases} 0 \\ 0.2498 \end{cases}$$

where the upper value after the brace is for $\sigma = 0$, and the lower value is for $\sigma = 1$, as shown for x_1. Computational results, including $\partial\rho/\partial\tilde{s}$ and $\partial V/\partial\tilde{s}$, are

$$\frac{1}{\rho_2}\left(\frac{\partial\rho}{\partial\tilde{s}}\right)_2 = \begin{cases} -0.5278 \\ -0.4395 \end{cases}, \quad \frac{1}{V_2}\left(\frac{\partial V}{\partial\tilde{s}}\right)_2 = \begin{cases} 0.1601 \\ 0.1333 \end{cases}$$

$$\frac{\partial\theta}{\partial s} = -0.1137$$

$$\frac{\partial\theta}{\partial\tilde{s}} = \begin{cases} -0.2433 \\ -0.1634 \end{cases}$$

$$\frac{\partial\theta}{\partial\tilde{n}} = \begin{cases} 0.3677 \\ 0.05631 \end{cases}$$

$$\frac{\partial\theta}{\partial n} = \begin{cases} -0.4260 \\ -0.1053 \end{cases}$$

It is evident from the ρ and V derivatives that the flow is expansive at state 2, where the shock is a weak solution shock. There is one Thomas point, for both the $\sigma = 0$ and 1 cases, on the shock at a β value larger than 41.81° (see Problem 24). When $\sigma = 0$, there is a Crocco point between the sonic and detachment states. Because $G_{cp} = 199.8$, one can show, for the axisymmetric case, that a Crocco point exists at a point on the shock downstream of where β is 41.81°.

As expected, $\partial\theta/\partial s$ has a negative value for the convex shock. Since n and $\tilde{n}$ have opposite orientations, the signs of $\partial\theta/\partial n$ and $\partial\theta/\partial\tilde{n}$ differ. The negative value for $\partial\theta/\partial\tilde{s}$ means the streamline curves downward, away from the shock. The substantial difference in $\partial\theta/\partial\tilde{n}$ and $\partial\theta/\partial n$ between the two-dimensional and axisymmetric shocks is largely due to the 0.25 value of $\sin\theta/y$, which, in the relevant equations, has been replaced with Equation (5.27).

6 Vorticity and Its Substantial Derivative

6.1 PRELIMINARY REMARKS

A systematic study is presented of the vorticity and its substantial derivative, both evaluated just downstream of a curved shock wave. As we know, the substantial derivative provides the rate of change of a property following a fluid particle. In a steady flow, this becomes the rate of change of a property along a streamline. The substantial derivative of the vorticity determines whether or not its strength is increasing or decreasing in the flow region just downstream of the shock.

Results and the analytical method may be of interest to researchers studying vortex-shock interaction, the external effect of vorticity on a boundary layer, or a detonation wave with cellular structure (see Problem 18).

The equations derived for ω and $D\omega/Dt$ are exact. They are also algebraic, explicit, and can readily be evaluated with a computer. The vorticity equation has the form of a shock jump condition in which the upstream flow is uniform and steady. Even though the analysis is in a flow plane and the Euler equations are not required, the vorticity result is limited to a two-dimensional or axisymmetric flow. The two vorticity parameters depend on γ, M_1, the slope of the shock, β, and its curvature, $-\beta'$. In the axisymmetric case there is also a dependence on the $\cos\beta/y$ curvature.

Starting in Section 6.4, application is for a detached bow shock, although the theory also applies to an attached shock.

Additional material can be found in Emanuel (2007) and Emanuel and Hekiri (2007). In the first of these references, general equations are obtained for $D\omega/Dt$ and for Crocco's equation in a diffusive, reacting, viscous, general gas mixture. Results for ω and $D\omega/Dt$, for a generic shock shape, are provided in the second reference and are repeated here, but with $D\omega/Dt$ corrected. (The author gratefully acknowledges the assistance of Hekiri for the recomputation and for the revised figure.)

The first two sections respectively derive equations for ω and $D\omega/Dt$. Subsequent sections provide analysis and parametric results for two generic shock shapes.

6.2 VORTICITY

The curl of $\vec{V}$ is normal to the streamlines in a two-dimensional or axisymmetric flow and therefore is tangent to the shock. Consequently, $\vec{\omega}$ is normal to the flow plane (i.e., it is proportional to $\hat{b}$). We can now write

$$\vec{\omega} = \omega \hat{b}, \quad \vec{V} = u\hat{t} + v\hat{n} \tag{6.1a,b}$$

In general, Equation (6.1a) does not hold in a three-dimensional flow. To kinematically demonstrate this in a simple way, use a Cartesian coordinate system where u, v, w are the x, y, z velocity components. Let u and v be functions only of x and y and let w be a nonzero constant (i.e., this is a two-dimensional flow with sweep) (Emanuel 2001, Chapter 10). A $\vec{\omega} \cdot \vec{V}$ calculation shows that the vorticity is not normal to the flow plane, which is angled with respect to the x,y plane. (This sweep flow model is also invoked in Section 9.5.)

It is convenient to introduce Crocco's equation:

$$\frac{\partial \vec{V}}{\partial t} + \vec{\omega} \times \vec{V} = T\nabla S - \nabla h_o \tag{6.2a}$$

where S is the entropy, and h_o is the stagnation enthalpy. Under the assumptions of steady, homenergetic flow, this reduces to

$$\vec{\omega} \times \vec{V} = T\nabla S \tag{6.2b}$$

which becomes

$$\omega = \frac{T}{u}\frac{\partial S}{\partial n}, \quad \omega = -\frac{T}{v}\frac{\partial S}{\partial s} \tag{6.3a,b}$$

By eliminating ω, the streamline isentropic equation is obtained. The entropy of a perfect gas is written as

$$S = S_o + \frac{R}{\gamma - 1} ln\left(\frac{p}{\rho^\gamma}\right) \tag{6.4}$$

where S_o is a constant, and R is the gas constant. The shock arc length derivative yields

$$\frac{\partial S}{\partial s} = \frac{R}{\gamma - 1}\left(\frac{1}{p}\frac{\partial p}{\partial s} - \frac{\gamma}{\rho}\frac{\partial \rho}{\partial s}\right) \tag{6.5a}$$

With the aid of Appendix E, this becomes

$$\frac{\partial S}{\partial s} = \gamma R \frac{Z^2 \beta'}{XY \tan\beta} \tag{6.5b}$$

At a normal shock, when β' is not zero, $(\partial S/\partial s)_2$ is infinite due to the $\tan\beta$ factor.

The vorticity is normalized by V_1 divided by an arbitrary reference length. By combining Equations (6.3b) and (6.5b) and replacing the temperature with $p/(\rho R)$, the desired result is obtained:

$$\omega_2 = -\frac{2}{\gamma+1}\frac{Z^2}{wX}\beta'\cos\beta \tag{6.6}$$

The vorticity and β' have opposite signs when β is in the first quadrant. For a convex shock, $(\partial S/\partial s)_2$ is negative, and, from Equation (6.3b), ω_2 is positive. This conclusion is in accord with Equation (6.6) when β' is negative and β is in the first quadrant. Since $\hat{b}$ points into the page, a positive $\overline{\omega}_2$ also points into the page. (For a quite different type of vorticity derivation, see Kanwal 1958a.)

Note that Equation (6.6) is independent of σ and that normal derivatives are not utilized in its derivation. The vorticity is zero when

$$\text{(i) } w = 1 \tag{6.7a}$$

$$\text{(ii) } \beta = 90° \tag{6.7b}$$

$$\text{(iii) } \beta' = 0 \tag{6.7c}$$

Condition (i) corresponds to the shock becoming a Mach wave, which, by itself, does not generate vorticity. The second condition occurs when the shock is normal to the freestream velocity. Condition (iii) is for a straight planar or conical shock.

It is useful to reexamine Equation (4.9b) by rederiving the vorticity equation by starting with the curl of the velocity:

$$\overline{\omega} = \nabla \times \overline{V} = \frac{1}{h_1 h_2 h_3}\begin{vmatrix} h_1\hat{t} & h_2\hat{n} & h_3\hat{b} \\ \dfrac{\partial}{\partial\xi_1} & \dfrac{\partial}{\partial\xi_2} & \dfrac{\partial}{\partial\xi_3} \\ h_1 u & h_2 v & 0 \end{vmatrix}$$

$$= \frac{1}{h_1 h_2 h_3}\left[h_2\frac{\partial}{\partial\xi_3}(h_1 u)\hat{n} + h_3\frac{\partial}{\partial\xi_1}(h_2 v)\hat{b} - h_3\frac{\partial}{\partial\xi_2}(h_1 u)\hat{b} \right.$$

$$\left. - h_1\frac{\partial}{\partial\xi_3}(h_3 v)\hat{t} \right] \tag{6.8a}$$

where Equation (6.1b) is utilized. The zero scalar values in the ξ_3 direction

$$\frac{\partial(h_2 v)}{\partial\xi_3} = 0, \quad \frac{\partial(h_1 u)}{\partial\xi_3} = 0$$

result in

$$\vec{\omega} = -\left(\frac{\partial u}{\partial n} - \frac{\partial v}{\partial s} + \frac{u}{h_1 h_2}\frac{\partial h_1}{\partial \xi_2} - \frac{v}{h_1 h_2}\frac{\partial h_2}{\partial \xi_1}\right)\hat{b} \tag{6.8b}$$

which verifies Equation (6.1a). With the aid of Equations (3.28), (3.37), and (3.35), we have

$$\vec{\omega}_2 = -\left[\frac{\partial u}{\partial n} - \frac{\partial v}{\partial s} + \beta'(u + v\tan\beta)\right]\hat{b} \tag{6.8c}$$

Appendix E now yields

$$\omega_2 = -\frac{2}{\gamma+1}\frac{\beta'\cos\beta}{wX}\left(Z^2 + X^2\tan^2\beta\right) \tag{6.8d}$$

which agrees with Equation (6.6) except for the $X^2\tan^2\beta$ term. This term stems from the $v\tan\beta$ term in Equation (6.8c), which in turn stems from a nonzero value for the curvature, at the shock, of the n-coordinate in the flow plane (i.e., $\kappa_o \neq 0$). One can show, in general, that

$$\kappa_o = \frac{1}{h_1 h_2}\frac{\partial h_2}{\partial \xi_1}$$

which becomes, at the shock,

$$\kappa_o = -\frac{f' f''}{\left(1 + f'^2\right)^{3/2}} = -\beta'\tan\beta$$

This also equals the negative value of $\hat{t}\cdot\partial\hat{n}/\partial n$. The $X^2\tan^2\beta$ term is in error; it results in an infinite value for ω_2 when the shock is normal to the freestream. From symmetry considerations, just downstream of the normal part of a detached shock, it is physically apparent that ω_2 must be zero, as is the case with Equation (6.6). The difficulty stems from using a correct three-dimensional formula, Equation (6.8a), when a surface analysis is required. This is apparent from the use of the entropy, Equation (6.4), which requires a shock. The error occurs once the equations in Equation (3.35) are introduced into Equation (6.8b), because the equations in Equation (3.35) do not yield $\kappa_o = 0$ at the shock's surface.

Problem 7 evaluates the acceleration in the flow plane at state 2. The $v\tan\beta$ factor also appears in part (b) of the problem and, unless deleted, results in an infinite acceleration for a normal shock with curvature.

There is no jump in the vorticity between its shock-generated value and a point in the flow infinitesimally downstream of the shock. In the next section, the jump phenomenon does not occur, because the derivation, starting with Equation (6.9), is not limited to just the downstream side of a shock. Appendix E is not utilized until Equations (6.14a) and (6.15a), and there is no $v\tan\beta$ term.

6.3 SUBSTANTIAL DERIVATIVE OF THE VORTICITY

Starting with the inviscid momentum equation, in vector form, one can show that (Emanuel 2001, Problem 4.5)

$$\frac{D\vec{\omega}}{Dt} = \frac{1}{\rho^2}\nabla\rho\times\nabla p + \vec{\omega}\cdot\left(\nabla\vec{V}\right) - \left(\nabla\cdot\vec{V}\right)\vec{\omega} \tag{6.9}$$

where $\vec{\omega}\cdot\left(\nabla\vec{V}\right)$ is zero in a two-dimensional flow but not in an axisymmetric one. The $\nabla\rho\times\nabla p$ term is referred to as a *barotropic term*; it is zero when $p = p(\rho)$. (Downstream of a curved shock, $p = p(\rho,S)$.)

The dyadic is first evaluated

$$\nabla\vec{V} = \left(\hat{t}\frac{\partial}{\partial s} + \hat{b}\frac{\partial}{\partial b} + \hat{n}\frac{\partial}{\partial n}\right)\left(u\hat{t} + v\hat{n}\right)$$

$$= \frac{\partial u}{\partial s}\hat{t}\hat{t} + u\hat{t}\frac{\partial\hat{t}}{\partial s} + \frac{\partial v}{\partial s}\hat{t}\hat{n} + v\hat{t}\frac{\partial\hat{n}}{\partial s} + \frac{\partial u}{\partial b}\hat{b}\hat{t}$$

$$+ u\hat{b}\frac{\partial\hat{t}}{\partial b} + \frac{\partial v}{\partial b}\hat{b}\hat{n} + v\hat{b}\frac{\partial\hat{n}}{\partial b} + \frac{\partial u}{\partial n}\hat{n}\hat{t} + u\hat{n}\frac{\partial\hat{t}}{\partial n} + \frac{\partial v}{\partial n}\hat{n}\hat{n} + v\hat{n}\frac{\partial\hat{n}}{\partial n} \tag{6.10a}$$

Note that

$$\frac{\partial u}{\partial b} = 0, \qquad \frac{\partial v}{\partial b} = 0$$

and with the equations in Equation (3.43), the dyadic becomes

$$\nabla\vec{V} = \left(\frac{\partial u}{\partial s} + \beta'v\right)\hat{t}\hat{t} + \left(\frac{\partial v}{\partial n} - u\beta'\tan\beta\right)\hat{n}\hat{n}$$

$$+ \frac{\sigma}{y}\left(u\sin\beta - v\cos\beta\right)\hat{b}\hat{b} + \left(\frac{\partial v}{\partial s} - \beta'u\right)\hat{t}\hat{n} + \left(\frac{\partial u}{\partial n} + v\beta'\tan\beta\right)\hat{n}\hat{t} \tag{6.10b}$$

where R, for consistency with Appendix E, is replaced with y. Since $\vec{\omega}$ is proportional to $\hat{b}$, the relatively simple result is obtained:

$$\vec{\omega}\cdot\left(\nabla\vec{V}\right) = \frac{\sigma}{y}\left(u\sin\beta - v\cos\beta\right)\omega\hat{b} \tag{6.11}$$

In a similar manner, the divergence term is

$$\nabla \cdot \vec{V} = \frac{\partial u}{\partial s} + \frac{\partial v}{\partial n} + \left(v - u \tan\beta\right)\beta' + \frac{\sigma}{y}\left(u \sin\beta - v \cos\beta\right) \tag{6.12}$$

We thereby obtain

$$\vec{\omega} \cdot \left(\nabla \vec{V}\right) - \left(\nabla \cdot \vec{V}\right)\vec{\omega} = -\left[\frac{\partial u}{\partial s} + \frac{\partial v}{\partial n} + \left(v - u \tan\beta\right)\beta'\right]\omega\hat{b} \tag{6.13}$$

which, with Appendix E, becomes

$$\vec{\omega} \cdot \left(\nabla \vec{V}\right) - \left(\nabla \cdot \vec{V}\right)\vec{\omega} = 2\sin\beta\left[\beta' - \frac{\beta'}{2wZ}G_2 + \frac{2}{\left(\gamma+1\right)^2}\frac{Y}{w}\frac{\sigma\cos\beta}{y}\right]\omega\hat{b} \tag{6.14a}$$

where

$$G_2 = g_2 - \frac{2}{\gamma+1}m\left(1+3w\right) + \frac{2}{\gamma+1}XZ \tag{6.14b}$$

The barotropic term is given by

$$\nabla\rho \times \nabla p = \left(\frac{\partial\rho}{\partial n}\frac{\partial p}{\partial s} - \frac{\partial\rho}{\partial s}\frac{\partial p}{\partial n}\right)\hat{b}$$

where the derivatives of ρ and p with respect to b are zero. This term becomes

$$\frac{1}{\rho^2}\nabla\rho \times \nabla p = \frac{16}{\left(\gamma+1\right)^3}\frac{\sin\beta\cos\beta}{w^2}\left[\frac{\left(\beta'\right)^2}{ZX}G_1 + \frac{\gamma-1}{2}Z^2\beta'\frac{\sigma\cos\beta}{y}\right]\hat{b} \tag{6.15a}$$

where

$$G_1 = w\left(mg_3 + g_4\right) - \frac{\gamma+1}{4}\left(mg_5 + g_6\right) \tag{6.15b}$$

The substantial derivative is finally

$$\left(\frac{D\omega}{Dt}\right)_2 = \frac{16}{\left(\gamma+1\right)^3}\frac{\sin\beta\cos\beta}{w^2}\left[\frac{\left(\beta'\right)^2}{XZ}G_1 + \frac{\gamma-1}{2}Z^2\beta'\frac{\sigma\cos\beta}{y}\right]$$

$$+ 2\sin\beta\left[\beta' - \frac{\beta'}{2wZ}G_2 + \frac{2}{\left(\gamma+1\right)^2}\frac{Y}{w}\frac{\sigma\cos\beta}{y}\right]\omega \tag{6.16}$$

When the shock is normal to the freestream, both ω_2 and $(D\omega/Dt)_2$ are zero. Far downstream, where β' goes to zero, both ω_2 and $(D\omega/Dt)_2$ also go to zero. For a blunt-body flow with a convex shock, both parameters have one extremum value.

6.4 GENERIC SHOCK SHAPE

Specific results utilize Billig's (1967) shock wave formula. It is a convenient, generic approach that enables various trends to be discerned for the two body shapes considered. Two-dimensional and axisymmetric nondimensional results are compared for the same value of the arc length along the shock, measured from where it is a normal shock. The Billig formula is for cylinder-wedge and sphere-cone bodies. Experimental air data were used to generate the shock shape and it is shown to be accurate, at least in the freestream Mach number range of interest (Billig 1967):

$$2 \le M_1 \le 6$$

In nondimensional form, the shape of the shock depends only on σ, M_1, and θ_b, where θ_b is the half angle of the wedge or cone. The parameter space for this study is

$$\gamma = 1.4, \quad \sigma = 0,1$$

$$M_1 = 2,4,6$$

$$\theta_b = 5°,10°,15°$$

There are thus nine two-dimensional and nine axisymmetric cases.

Various lengths and angles are defined in Figure 6.1. An overbar (not shown in the figure) denotes a dimensional quantity. Lengths are normalized by the radius of the cylinder or sphere—that is,

$$x = \frac{\overline{x}}{\overline{R}_b}, \quad y = \frac{\overline{y}}{\overline{R}_b}, \quad s = \frac{\overline{s}}{\overline{R}_b}, \quad n = \frac{\overline{n}}{\overline{R}_b}, \quad \Delta = \frac{\overline{\Delta}}{\overline{R}_b}, \quad r = \frac{\overline{R}_s}{\overline{R}_b} \tag{6.17}$$

where $\overline{n}$ is distance normal to the shock in the downstream direction, and $\overline{\Delta}$ is the shock stand-off distance.

Billig (1967) provides an empirical relation for the shock shape:

$$x = -1 - \Delta + r \frac{\left(1 + y^2 \dfrac{\tan^2 \beta_\infty}{r^2}\right)^{1/2} - 1}{\tan^2 \beta_\infty} \tag{6.18}$$

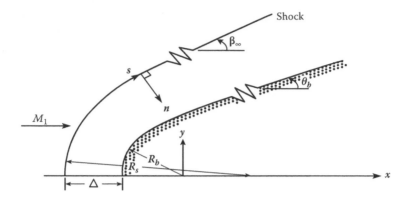

FIGURE 6.1   Body and shock wave sketch showing various lengths and the β_∞ and θ_b angles.

where

$$\Delta = \begin{cases} 0.386 \exp(4.67 / M_1^2), & \sigma = 0 \\ 0.143 \exp(3.24 / M_1^2), & \sigma = 1 \end{cases} \tag{6.19a}$$

$$r = \begin{cases} 1.386 \exp\left[1.8/(M_1 - 1)^{0.75}\right], & \sigma = 0 \\ 1.143 \exp\left[0.54/(M_1 - 1)^{1/2}\right], & \sigma = 1 \end{cases} \tag{6.19b}$$

Problem 20 provides an analytic estimate, using Appendix E, for the shock stand-off distance, which can then be compared with Equation (6.19a).

It is analytically convenient to introduce

$$z = 1 + \Delta + x \tag{6.20}$$

where z is zero at the location where the shock intersects the axis of symmetry. Equation (6.18) is inverted, with the simple result

$$y = \left(2rz + z^2 \tan^2 \beta_\infty\right)^{1/2} \tag{6.21}$$

where this quadratic can be shown to be a hyperbola ($\sigma = 0$) or hyperboloid ($\sigma = 1$) shock.

Far downstream, the shock angle β_∞ is for a sharp cone or wedge with the same half angle, θ_b. If θ_b is zero, β_∞ equals the freestream Mach angle. For a wedge, β_∞ and θ_b are related by Equation (2.28) or Appendix B, which provides β_∞ given θ_b. Table 6.1 tabulates results for β_∞, where the axisymmetric results are provided by the Ames Research Staff (1953, p. 48).

TABLE 6.1
β_∞ versus θ_b, in Degrees, for a Wedge ($\sigma = 0$) and a Cone ($\sigma = 1$)

	$M_1 = 2$		$M_1 = 4$		$M_1 = 6$	
θ_b	$\sigma = 0$	$\sigma = 1$	$\sigma = 0$	$\sigma = 1$	$\sigma = 0$	$\sigma = 1$
5	34.302	30.15	18.021	15.0	13.160	10.6
10	39.314	31.15	22.234	17.65	17.587	14.5
15	45.344	33.9	27.063	21.9	22.672	19.0

6.5 SLOPE, CURVATURE, ARC LENGTH, AND SONIC POINT

Equation (6.21) is differentiated, to yield

$$\frac{dy}{dx} = \frac{r + z \tan^2 \beta_\infty}{y} \tag{6.22}$$

or, for the slope,

$$\beta = \tan^{-1}\left(\frac{r + z \tan^2 \beta_\infty}{y}\right) \tag{6.23}$$

The derivative of β

$$\frac{d\beta}{dx} = -\frac{2}{y\left[y^2 + \left(r + z \tan^2 \beta_\infty\right)^2\right]} \tag{6.24}$$

is required for the curvature, which is

$$\beta' = \frac{d\beta}{ds} = \frac{d\beta}{dx}\frac{dx}{ds} \tag{6.25}$$

where

$$\frac{ds}{dx} = \left[1 + \left(\frac{dy}{dx}\right)^2\right]^{1/2} = \left(1 + \tan^2 \beta\right)^{1/2} = \frac{1}{y}\left[y^2 + \left(r + z \tan^2 \beta_\infty\right)^2\right]^{1/2} \tag{6.26}$$

With the aid of Equation (6.24), the negative of the curvature becomes

$$\beta' = -\frac{r^2}{\left[y^2 + \left(r + z \tan^2 \beta_\infty\right)^2\right]^{3/2}} \tag{6.27}$$

When $z = 0$, the following is readily obtained:

$$x = -1 - \Delta, \quad y = 0, \quad s = 0, \quad \beta = 90°, \quad \beta' = -\frac{1}{r} \tag{6.28}$$

where r is the normalized radius of curvature of the shock at its nose. Observe that $-\beta'$ has its maximum value at $z = 0$ and decreases toward zero as z, or s, becomes infinite.

The arc length stems from the integration of Equation (6.26)—that is,

$$s = \int_0^z \left[y^2 + \left(r + z \tan^2 \beta_\infty \right)^2 \right]^{1/2} \frac{dz}{y} \tag{6.29a}$$

which is written as

$$s = \frac{1}{\cos \beta_\infty} \int_0^z \left(\frac{r^2 \cos^2 \beta_\infty + 2rz + z^2 \tan^2 \beta_\infty}{2rz + z^2 \tan^2 \beta_\infty} \right)^{1/2} dz \tag{6.29b}$$

With the substitution

$$t = 1 + z \frac{\tan^2 \beta_\infty}{r}, \quad z = \frac{r}{\tan^2 \beta_\infty}(t - 1) \tag{6.30}$$

the integrand has the standard form

$$s = \frac{r}{\cos \beta_\infty \tan^2 \beta_\infty} \int_0^t \left(\frac{t^2 - \cos^2 \beta_\infty}{t^2 - 1} \right)^{1/2} dt \tag{6.29c}$$

Gradshteyn and Ryzhik (1980) provide the integral in terms of elliptic integrals, with the result

$$s = \frac{r}{\cos \beta_\infty \tan^2 \beta_\infty} \left[\sin^2 \beta_\infty F\left(\varphi \backslash \frac{\pi}{2} - \beta_\infty \right) \right.$$

$$\left. - E\left(\varphi \backslash \frac{\pi}{2} - \beta_\infty \right) + t \left(\frac{t^2 - 1}{t^2 - \cos^2 \beta_\infty} \right)^{1/2} \right] \tag{6.29d}$$

where t is given by Equation (6.30) and

$$\varphi = \sin^{-1} \left(\frac{t^2 - 1}{t^2 - \cos^2 \beta_\infty} \right)^{1/2} \tag{6.31}$$

The standard notation of Milne-Thomson (1972) is used for the first and second elliptic integrals, $F(\varphi \backslash \alpha)$ and $E(\varphi \backslash \alpha)$, respectively, instead of that in Gradshteyn and Ryzhik (1980).

The sonic point location β^* is provided by Equation (5.14) and is independent of σ and θ_b. This location, of interest when discussing results, determines the computational spacing, which ranges from $z = 0$ to $z = 19z^*$. The spacing clusters points near $z = 0$ and include z^*. To obtain z^*, solve Equation (6.23) for z, replace y with Equation (6.21), and set $\beta = \beta^*$, with the result

$$z^* = \frac{r}{\tan^2 \beta_\infty} \left[\frac{\tan \beta^*}{\left(\tan^2 \beta^* - \tan^2 \beta_\infty \right)^{1/2}} - 1 \right] \qquad (6.32)$$

6.6　RESULTS

In accord with Equation (6.6), ω depends linearly on β' and not directly on σ. (Again, for notational convenience, the subscript 2 is suppressed.) It does, however, indirectly depend on σ through β_∞, Δ, and r. By differentiating Equation (6.6) with respect to s, one can show that ω has an extremum value when

$$\frac{\beta''}{\left(\beta' \right)^2} = \tan \beta - \frac{2\left(1 + \gamma w \right)}{XZ \tan \beta} \qquad (6.33)$$

is satisfied. This implicit equation provides the β value where ω is a maximum (see Figure 6.2), where w, X, and Z depend on β through $\sin^2 \beta$.

In Figures 6.2 through 6.4, the solid (dashed) curves are for $\sigma = 0$ (1). As shown in Figure 6.2, the maximum of ω increases with M_1. At small s, the axisymmetric ω value substantially exceeds its two-dimensional counterpart. The upper s limit for this is roughly 2, where the curves tend to cross. The figure demonstrates a weak dependence on θ_b, but a strong dependence on dimensionality and M_1. The weak θ_b dependence, for the values chosen, holds throughout this study. On the other hand, the dimensionality dependence gradually weakens as M_1 increases. For instance, when $M_1 = 2$, the axisymmetric ω value can exceed its two-dimensional counterpart by an order of magnitude. When $M_1 = 6$, the difference is less than a factor of 2.

For large s, the axisymmetric ω value decays more rapidly and the curves cross. Table 6.2 shows the sonic value, s^*, for the 18 cases. Except for the $\sigma = 0$, $M_1 = 2$ cases, the s^* values fall between 0.65 and 1.5. By comparing Figure 6.2 and Table 6.2, observe that the peak ω values occur at a state that ranges from subsonic to low supersonic.

The strong dependence on dimensionality and M_1, and the weakening of the dimensionality dependence with increasing M_1, stems directly from the variation of the curvature with s (Figure 6.3). On the symmetry axis, the $|\beta'|$ value is significantly larger when $\sigma = 1$ than its $\sigma = 0$ counterpart, especially when $M_1 - 1$ is small. Also apparent is the weak dependence of β' on θ_b and the closeness of the curves when s exceeds 4. The dimensionality difference is caused by the shock being closer to the body when $\sigma = 1$. This is evident in Table 6.2, which shows r, the $\bar{R}_s / \bar{R}_b$ ratio. (Although listed when θ_b is 5°, r is independent of θ_b.) At a low freestream Mach

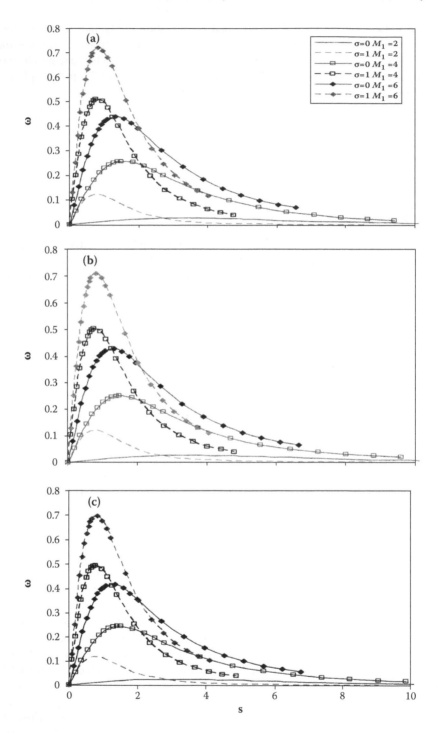

FIGURE 6.2   Vorticity (a) $\theta_b = 5°$, (b) $\theta_b = 10°$, and (c) $\theta_b = 15°$.

TABLE 6.2
Location of the Sonic State and the r Parameter

σ	M_1	θ_b	s^*	r
0	2	5	5.150	8.385
		10	5.348	
		15	5.749	
	4	5	1.472	3.053
		10	1.483	
		15	1.499	
	6	5	1.066	2.374
		10	1.071	
		15	1.078	
1	2	5	1.178	1.961
		10	1.184	
		15	1.202	
	4	5	0.7501	1.561
		10	0.7526	
		15	0.7577	
	6	5	0.6520	1.455
		10	0.6540	
		15	0.6573	

number, the difference between the two-dimensional and axisymmetric shock stand-off distances is quite large. At small s and small $M_1 - 1$ values, because of the three-dimensional relief effect associated with an axisymmetric flow, β' strongly depends on dimensionality. This dependence weakens as M_1 increases and the shock stand-off distance rapidly decreases for a two-dimensional shock. The three-dimensional relief effect is therefore responsible for the dimensionality difference in the vorticity and its decline with increasing Mach number.

Examination of $(\partial p/\partial n)_2$, $(\partial \rho/\partial n)_2$, and $(\partial T/\partial n)_2$ in Appendix E.3 shows that their axisymmetric σ terms are positive and compressive. These terms, however, are more than offset by the β' terms. (Remember that $\cos\beta/y$ is finite when $y \to 0$.) The β' terms experience a large increase in magnitude when s is small. In this circumstance, there is a substantial isentropic expansion just downstream of an axisymmetric shock but slightly removed from the centerline.

The relief effect is evident in the results of Problem 24, where the Thomas point $[(\partial p/\partial \tilde{s}) = 0]$ has a y value of 2.591 when $\sigma = 0$ and 0.9297 when $\sigma = 1$. The flow therefore becomes expansive much closer to the symmetry line when the flow is axisymmetric. This relatively close and intense expansion, when $\sigma = 1$, represents the relief effect.

In contrast to ω, $D\omega/Dt$ is proportional to $(\beta')^2$ when $\sigma = 0$. When $\sigma = 1$, $D\omega/Dt$ is also proportional to a $\beta'(\cos\beta/y)$ term. On the symmetry axis, Equation (5.2) provides $-r^{-1}$ for β', while $\beta'(\cos\beta/y)$ equals $-r^{-2}$. Nevertheless, Equation (6.16) shows that $D\omega/Dt$ is zero on the symmetry axis for both two-dimensional and axisymmetric shocks. As s becomes large, both $(\beta')^2$ and $\beta'(\cos\beta/y)$ approach zero, as does $D\omega/Dt$, as evident in Figure 6.4.

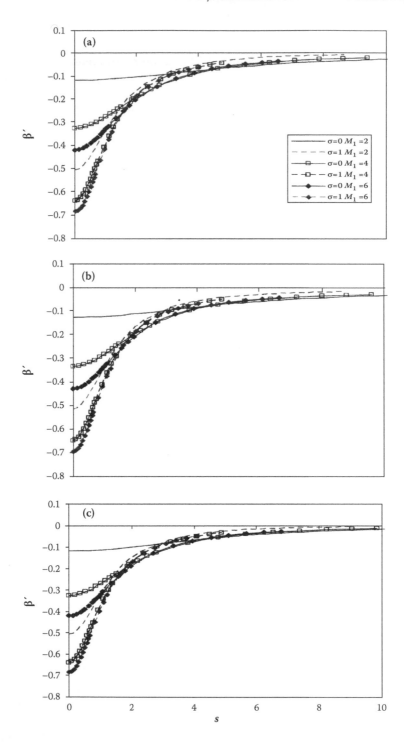

FIGURE 6.3   Shock curvature (a) $\theta_b = 5°$, (b) $\theta_b = 10°$, and (c) $\theta_b = 15°$.

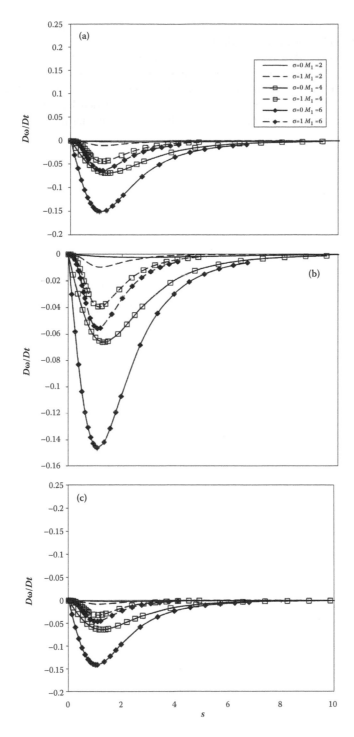

FIGURE 6.4 Substantial derivative of the vorticity (a) $\theta_b = 5°$, (b) $\theta_b = 10°$, and (c) $\theta_b = 15°$.

Except when $M_1 = 2$, both ω and $D\omega/Dt$ have extremum near $s = 1$. While ω is positive, $D\omega/Dt$ is negative and the magnitude of the vorticity therefore decreases in value (i.e., decays) downstream of the shock. The $D\omega/Dt$ value for $M_1 = 2$, $\sigma = 0$ is barely visible; the axisymmetric case decays faster. This reverses when $M_1 = 4$ and 6. When $M_1 = 4$, the two-dimensional case decays about twice as fast as the axisymmetric case. The margin is even bigger in favor of $\sigma = 0$ when $M_1 = 6$. As with vorticity, the magnitude of $D\omega/Dt$ increases rapidly with M_1 and hardly changes with θ_b.

7 Shock Wave Triple-Point Morphology

7.1 PRELIMINARY REMARKS

Triple points occur in steady and unsteady supersonic flows, such as transonic flow over an airfoil or in a jet emanating from a supersonic nozzle. A triple point also occurs when an incident shock is unable to regularly reflect from a wall, a symmetry line (in an axisymmetric flow), or a symmetry plane. At a triple point, really line, three shocks intersect: an incident (I) shock, a reflected (R) shock, and a Mach stem (M). At the intersection, a slipstream (SS), which is a free shear layer in a viscous analysis, is generated. The slipstream is defined by two conditions: the pressure is the same across it, and the velocities on each side are tangent to it. These constraints are the basis for any local triple-point analysis. Triple points are usually discussed within the context of shock wave reflection phenomena (Azevedo and Liu 1993; Ben-Dor 2007; Courant and Friedrichs 1948; Henderson and Menikoff 1998; Hornung 1986; Ivanov et al. 1998; Kalghatgi and Hunt 1975; Mouton and Hornung 2007; Uskov and Chernyshov 2006; Uskov and Mostovykh 2011). The focus here, however, is on triple-point morphology, and not on the reflection process.

Shock/shock interference (for example, see Borovoy et al. 1997; Edney 1968) occurs in a supersonic or hypersonic flow when an upstream shock impinges on a detached bow shock. There are six types of interaction (see Figure 1 in Borovoy et al. 1997). In two of these, the two shocks cross or coalesce (with a centered expansion fan). In each of the other four interactions, there are two distinct triple points. In this type of flow, the wall reflection process is not germane, although it is convenient to retain the I, R, and M designations. This labeling, however, is now somewhat arbitrary. For instance, the two triple points are connected by a shock that can be viewed, for both triple points, as a Mach stem, or, in certain orientations, as a reflected shock for the upstream point and as the incident shock for the downstream one. One constant, however, is that the flow, just downstream of the incident shock, at the triple point, must be supersonic. As sketched in Figure 7.1, the flow elsewhere behind the incident shock need not be supersonic. The subsequent analysis also covers shock/shock interference triple points.

It has been customary to use an approach based on shock-polar diagrams whose coordinates correspond to the pressure and flow angle. A novel approach is introduced that is analytically/computationally straightforward and physically transparent. The pressure condition becomes a linear equation, while the tangency condition is a simple transcendental equation. As is known (Henderson 1964) for given values of γ, the upstream Mach number, M_1, and the incident shock wave angle, β_I, the number of triple-point solutions ranges from 0 to 3. In our approach, this array of

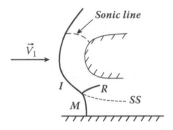

FIGURE 7.1   A curved, incident shock at a triple point.

possible solutions occurs within a parameter window that focuses the analytical/computational effort. A morphology of solutions are provided for $\gamma = 1$, 1.4, and 5/3. It has been asserted that the qualitative nature of the solutions does not vary with γ (Hornung 1986). As will be shown, this is not the case. To our knowledge, this study is the first to systematically investigate the impact of changes in the ratio of specific heats. While $\gamma = 1$ is not physically realistic, gases with a large number of atoms, some of which are heavy (e.g., UF_6) come close. Moreover, the equations considerably simplify in this limit.

The analysis assumes a perfect gas and the time-independent algebraic shock wave equations. It is local to the triple point and utilizes a single flow plane for all three shocks. The analysis, locally and at a given instant of time, thus holds for an unsteady, three-dimensional flow, including where the upstream flow is nonuniform. Questions of stability and hysteresis (Henderson and Menikoff 1998; Ivanov et al. 1998) are not considered. The stability analysis in Henderson and Menikoff is in terms of a convex equation of state and is not germane to a perfect gas analysis. The hysteresis analysis in Ivanov et al. is for a regular reflection/Mach reflection transition. Of the four types of interactions discussed in Henderson and Menikoff, the first three (Mach reflection, degenerate cross-node, degenerate overtake node) are covered. The degenerate overtake node corresponds to an inverted reflected shock, while the fourth type (two outgoing shocks) is not included, because it is a four-shock system.

A shock is locally characterized by γ, its upstream Mach number, and its wave angle β. This angle and the velocity turn angle, θ, are measured relative to the upstream velocity. It is analytically convenient, in this chapter, to consider these angles as limited to the first quadrant, regardless of the shock's orientation.

To add perspective to the subsequent analysis, a typical Mach reflection flow pattern is described, as sketched in Figure 7.2. The wedge is straight as is the incident shock. Region 2 is a uniform, supersonic flow that terminates at the reflected shock between the triple point (tp) and a' and along the leading edge characteristic a-a'. The part of the reflected shock between the triple point and point a' is drawn as straight; this will be modified shortly. There is a centered expansion that originates at the wedge's shoulder whose leading edge characteristic is denoted as a-a'-a''. This expansion is partly transmitted into the flow region downstream of R. Region 3 is triangular, whose vertices are the triple point and points a' and a''. This region is usually supersonic. Verification of this is provided later by Tables 7.3 and 7.4, which provide the Mach number, M_3, just downstream of R, at the triple point. The Mach

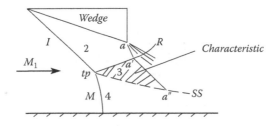

FIGURE 7.2 A conventional Mach reflection flow pattern.

stem is shown as concave relative to the upstream flow, and region 4 is subsonic. When M is concave, the subsequent analysis demonstrates that the slipstream is oriented downward, as pictured in the figure. Later, this is referred to as a type (b) or (c) triple-point configuration. Between the wall and SS, the subsonic flow experiences an expansion similar to what it would experience in a converging subsonic nozzle. Thus, the pressure along SS, from the triple point onward, at least to point a'', decreases. This decrease along SS generates an expansion wave that is transmitted via left-running characteristics, as sketched in the figure. The wave interacts with R and with the expansion fan. The interaction with R weakens it and causes R to slightly curve. The reflected shock between the triple point and a' is thus not straight. The interaction with the right-running characteristics of the expansion fan is weak.

The sketch in Figure 7.2 is not unique. A triple point may have a convex Mach stem, in which case the SS points upward. Later, this is referred to as a type (a) triple-point configuration. In addition, the flow downstream of R may be subsonic; that downstream of M may be supersonic.

The subsequent triple-point solution is purely algebraic; curvatures are not required, nor does the solution provide this information. Moreover, the triple-point curvatures of I, M, SS, and R are not independent of each other. Furthermore, if the wedge's surface is curved, then I as well as the other features generally have finite curvatures at the triple point. Using curved shock theory, Molder (2012) has developed an approach for obtaining the various curvatures and other gradients at the triple point.

Figure 7.3 is another Mach reflection triple-point sketch. The velocities and states, denoted by subscripts 1 through 4, apply only in the immediate vicinity of the triple point. The Mach stem is drawn as concave to the upstream flow. It may also be convex, in which case, β_M and θ_M have a first-quadrant, counterclockwise orientation. As indicated in the shock/shock interference discussion, the presence of a wall is not required. Instead of a concave (convex) Mach stem, the Mach stem need only slope in the downstream (upstream) direction when measured from the triple point. It is, nevertheless, convenient to refer to its orientation as convex or concave as if the triple point was part of a Mach reflection configuration. Remember, however, that curvatures are not part of the subsequent solution process. In all sketches, $\vec{V}_3$ and $\vec{V}_4$ are tangent to the slipstream. The incident shock in Figure 7.3 is drawn as a straight, weak-solution shock that could originate at the sharp leading edge of a wedge. This need not be the case. As sketched in Figure 7.1, I is part of a curved bow shock.

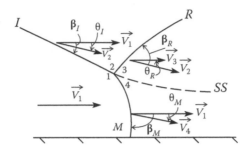

FIGURE 7.3   The β, θ angles and velocities at a triple point, with a concave Mach stem.

Of course, the Mach number, M_2, at the triple point, just downstream of I must be supersonic.

There are a number of transition, or special, cases. The first is when the Mach stem is neither convex nor concave but is a normal shock. A second case is where the reflected shock is a normal shock. A third case occurs when the reflected shock is inverted and its slope is in the upstream direction. Typically, the reflected shock is presumed to be weak, while the Mach stem is considered to be a strong solution shock. Another case occurs when one, or both, shock deviates from this pattern. The formulation does not require special provisions for any of these cases; they are discussed when appropriate.

Section 7.2 contains the triple-point analysis, while the method of solution is relegated to Section 7.3. Computational results are presented in Section 7.4 for a range of upstream Mach numbers and for three values for the ratio of specific heats. This chapter is based on the analysis by Hekiri and Emanuel (2011).

7.2 ANALYSIS

For the analysis, γ, M_1, and β_I are prescribed, where the wave angle of the incident shock, β_I, is bounded:

$$\beta_{I\mu} + \varepsilon \leq \beta_I \leq \beta_I^* - \varepsilon \tag{7.1}$$

where

$$\varepsilon = \text{small positive constant } (= 10^{-3} \text{ rad.}) \tag{7.2a}$$

$$\beta_{I\mu} = \sin^{-1}\left(\frac{1}{M_1}\right) \tag{7.2b}$$

and β_I^* is given by Equation (5.14). If β_I has its Mach wave value, $\beta_{I\mu}$, there can be no triple point. The upper bound ensures that M_2 is supersonic, not just sonic. Hence, R is also not a Mach wave. In the subsequent analysis, the limits, when $\varepsilon \to 0$, and when $\beta_I \to \beta_{I\mu}$ or $M_2 \to 1$ are discussed.

The pressure condition, $p_4 = p_3$, or

$$\frac{p_4}{p_1} = \frac{p_2}{p_1}\frac{p_3}{p_2} \tag{7.3a}$$

becomes, with the aid of the oblique shock equations,

$$x_M = ax_R + b \tag{7.3b}$$

where

$$x_i = \sin^2\beta_i, \qquad i = I, R, M \tag{7.4a}$$

$$w_i = M_i^2 \sin^2\beta_i \tag{7.4b}$$

$$X_i = 1 + \frac{\gamma-1}{2}w_i \tag{7.4c}$$

$$Y_i = \gamma w_i - \frac{\gamma-1}{2} \tag{7.4d}$$

$$Z_i = w_i - 1 \tag{7.4e}$$

$$M_I = M_1, \quad M_R = M_2, \quad M_M = M_1 \tag{7.4f}$$

$$a = \frac{2}{\gamma+1}\frac{X_I}{M_I^2}\left[1 + \left(\frac{\gamma+1}{2}\right)^2\frac{w_I\left(M_I^2 - w_I\right)}{X_I^2}\right] \tag{7.4g}$$

$$b = -\frac{\gamma-1}{\gamma+1}\frac{Z_I}{M_I^2} \tag{7.4h}$$

The analysis for a triple point is distinguished from other configurations when each of the pressure ratios in Equation (7.3a) is written in terms of an oblique shock equation.

With the above notation, M_2 is given by (see Appendix E.1)

$$M_2^2 = \frac{X_I}{Y_I}\left[1 + \left(\frac{\gamma+1}{2}\right)^2\frac{w_I\left(M_I^2 - w_I\right)}{X_I^2}\right] \tag{7.5}$$

Equation (7.3b) is a linear equation for the x_R and x_M unknowns. A second relation is provided by a velocity tangency constraint.

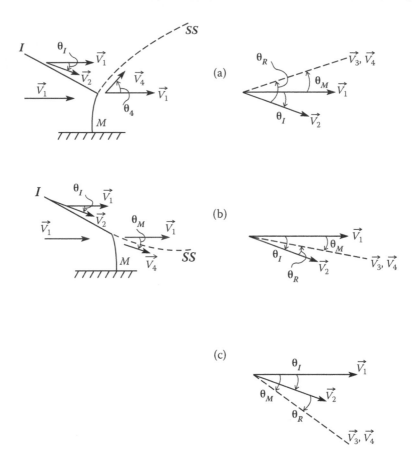

FIGURE 7.4 The three types of triple points. In (a) the Mach stem is convex, while in (b) and (c) it is concave. In (b) $\vec{V}_3$ and $\vec{V}_4$ are between $\vec{V}_1$ and $\vec{V}_2$, while in (c) $\vec{V}_3$ and $\vec{V}_4$ are clockwise from $\vec{V}_2$.

Figure 7.4 is a series of sketches illustrating the three possible tangency conditions. For purposes of clarity, the reflected shock is not shown. On the right side are sketches showing the various θ angles and the velocities, where the dashed line represents $\vec{V}_3$, $\vec{V}_4$, and SS. In type (a), the Mach stem is convex, relative to the upstream flow, SS slopes upward, and $\vec{V}_3$ and $\vec{V}_4$ are counterclockwise from $\vec{V}_1$. From the sketch on the right, the tangency condition is

$$\theta_I = \theta_R - \theta_M \tag{7.6a}$$

Types (b) and (c) have a concave Mach stem and SS slopes downward. As shown in the figure, SS is between $\vec{V}_1$ and $\vec{V}_2$ in type (b), while it is rotated clockwise from $\vec{V}_2$ in type (c). The (b) and (c) tangency conditions are

$$\theta_I = \theta_R + \theta_M \tag{7.6b}$$

$$\theta_I = \theta_M - \theta_R \qquad (7.6c)$$

respectively, and θ_i is given by

$$\tan \theta_i = \frac{1}{\tan \beta_i} \frac{M_i^2 \sin^2 \beta_i - 1}{1 + \left(\dfrac{\gamma+1}{2} \right) M_i^2 - M_i^2 \sin^2 \beta_i}, \qquad i = I, R, M \qquad (7.7)$$

Once γ, M_1, and β_I are prescribed, θ_I is known. With Equation (7.4a), θ_R and θ_M are written as

$$\theta_R = \tan^{-1} \left[\left(\frac{1 - x_R}{x_R} \right)^{1/2} \frac{M_2^2 x_R - 1}{1 + \left(\dfrac{\gamma+1}{2} \right) M_2^2 - M_2^2 x_R} \right] \qquad (7.8a)$$

$$\theta_M = \tan^{-1} \left[\left(\frac{1 - x_M}{x_M} \right)^{1/2} \frac{M_1^2 x_M - 1}{1 + \left(\dfrac{\gamma+1}{2} \right) M_1^2 - M_1^2 x_M} \right] \qquad (7.8b)$$

The desired x_R, x_M solution occurs when at least one of the following relations is satisfied:

$$F_a = 1 - \frac{\theta_R - \theta_M}{\theta_I} = 0 \qquad (7.9a)$$

$$F_b = 1 - \frac{\theta_R + \theta_M}{\theta_I} = 0 \qquad (7.9b)$$

$$F_c = 1 + \frac{\theta_R - \theta_M}{\theta_I} = 0 \qquad (7.9c)$$

As evident from the foregoing, a triple-point solution only requires the solution of algebraic equations.

The detachment wave angle for the Mach stem and the incident shock is

$$\sin^2 \beta_{Md} = \frac{\gamma+1}{4\gamma M_1^2} \left\{ M_1^2 - \frac{4}{\gamma+1} + \left[M_1^4 + 8 \left(\frac{\gamma-1}{\gamma+1} \right) M_1^2 + \frac{16}{\gamma+1} \right]^{1/2} \right\} \qquad (7.10)$$

By replacing M_1 with M_2, this relation also provides the reflected shock detachment wave angle, β_{Rd}. These detachment wave angles distinguish between the weak and strong solutions.

As defined, the β, θ angles are not always convenient for visualizing the orientation of the shocks and the slipstream. An overbar denotes an angle measured relative to $\vec{V}_1$, or the x-coordinate, in a counterclockwise orientation, as illustrated in the top panel of Figure 7.5. Equations for the barred angles are given in Table 7.1. Figure 7.5 contains sketches, based on the $\gamma = 1.4$, $M_1 = 6$ case, for the three solution types, where the type (c) inverted reflected wave and not inverted cases are shown. Because the reflected shock requires $\vec{V}_3$, relative to $\vec{V}_2$, to be rotated closer to R (see Figure 7.5), neither types (a) nor (b) can have an inverted reflected shock. The inverted/not inverted condition is

$$\bar{\beta}_R = \begin{cases} < 90°, & not\ inverted \\ > 90°, & inverted \end{cases} \qquad (7.11a)$$

and with Table 7.1, this becomes

$$\beta_R + \theta_I = \begin{cases} > 90°, & not\ inverted \\ < 90°, & inverted \end{cases} \qquad (7.11b)$$

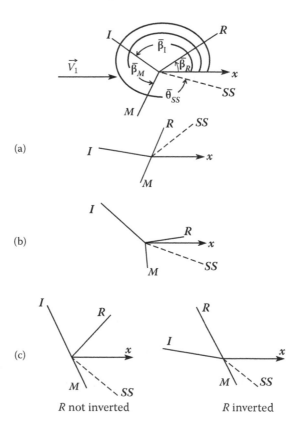

(a)

(b)

(c)

R not inverted   R inverted

FIGURE 7.5   The barred angles when $\gamma = 1.4$, $M_1 = 6$, where in (a) $\beta_I = 9.651°$, (b) $\beta_I = 43.77°$, (c) R not inverted $\beta_I = 66.52°$, R inverted $\beta_I = 9.651°$.

TABLE 7.1

Definitions of Barred Angles, Measured Counterclockwise from $\bar{V}_1$, in Degrees

	Type		
	(a)	**(b)**	**(c)**
$\bar{\beta}_I$	$180 - \beta_I$	$180 - \beta_I$	$180 - \beta_I$
$\bar{\beta}_R$	$\beta_R - \theta_I$	$\beta_R - \theta_I$	$180 - (\beta_R + \theta_I)$
$\bar{\beta}_M$	$180 + \beta_M$	$360 - \beta_M$	$360 - \beta_M$
$\bar{\theta}_{ss}$	θ_M	$360 - \theta_M$	$360 - \theta_M$

Finally, the R and M shocks are normal when

$$\beta_R = 90°, \qquad \bar{\beta}_M = 270° \qquad (7.12a,b)$$

respectively. Alternatively, $\bar{\theta}_{ss} = 0°$ when M is a normal shock.

7.3 SOLUTION METHOD

For a real-valued solution, the x_i must satisfy

$$\frac{1}{M_1^2} < x_I < \sin^2 \beta_I^*, \qquad \frac{1}{M_2^2} < x_R \le 1, \qquad \frac{1}{M_1^2} < x_M \le 1 \qquad (7.13a,b,c)$$

where the lower bounds represents a Mach wave, while the x_R, x_M upper bounds represent a normal shock. Relation (7.13a) is equivalent to Equation (7.1). The lower and upper bounds on x_R and x_M are required by the equations in Equation (7.8) in order that θ_R and θ_M have real, first-quadrant values.

Equation (7.13) represents the parameter window mentioned in the Preliminary Remarks. This window excludes any shock from being a Mach wave and requires that M_2 exceed unity. The closeness of a solution to one of the edges is partly governed by ε (Equation 7.2a). At the time this work was performed, the author was unaware of the analysis by Uskov and Chernyshov (2006), where solutions very close to this edge are investigated. In retrospect, a value of $\varepsilon = 10^{-5}$ rad. would have generated these solutions.

The algorithm starts with prescribed values for γ and M_1. An i-loop for β_I is established:

$$\Delta\beta_I = \frac{1}{N}\left(\beta_I^* - \beta_{I\mu} - 0.002\right) \qquad (7.14a)$$

$$\beta_{I,1} = \beta_{I\mu} + 0.001 \qquad (7.14b)$$

$$\beta_{I,i} = \beta_{I\mu} + (i - 1)\Delta\beta_I, \qquad i = 1, 2, ..., N \qquad (7.14c)$$

that spans the Equation (7.1) range, the angles are in radians, and the constants stem from Equation (7.2a). A second, inner k-loop for x_R utilizes

$$x_{R,1} = \frac{1}{M_2^2} + 10^{-4} \tag{7.15a}$$

$$x_{R,N+1} = \min\left(1, \frac{1-b}{a}\right) \tag{7.15b}$$

$$\Delta x_R = \frac{x_{R,N+1} - x_{R,1}}{N} \tag{7.15c}$$

$$x_{R,k} = x_{R,1} + (k-1)\,\Delta x_R, \qquad k = 1, 2, \ldots, N \tag{7.15d}$$

$$x_{M,k} = ax_{R,k} + b \tag{7.15e}$$

With Equation (7.4g,h), the demarcation between the two upper limits for $x_{R,N+1}$ is

$$\sin^2\beta_{Idem} = \frac{2}{\gamma+1}\frac{M_1^2 - 1}{M_1^2} \tag{7.16}$$

The $x_{R,N+1} = 1$ limit corresponds to a normal reflected shock, while the Mach stem is a normal shock in the $(1-b)/a$ case. In retrospect, a 10^{-5} value in Equation (7.15a) would be advisable.

The bounds of the equations in Equation (7.13) are adhered to in the above loops. For instance, if $x_R = M_2^{-2}$, Equation (7.3b) then yields $x_M = M_1^{-2}$, and all three disturbances are Mach waves. When R and M are normal shocks (see Equation 7.19), the incident shock is a Mach wave, and this point, on the window's edge, is excluded.

The two loops almost cover the window of possible triple-point solutions for given γ and M_1 values. The "almost" qualifier is unnecessary when the various limits, such as ε in Equation (7.2a), go to zero. As will become apparent, in this limit, the window is a necessary but not a sufficient condition for a solution.

In this study, N is set at 10, thereby generating an 11×11 (x_I, x_R) array of points for the window. (The x_M value is determined by Equation 7.15e.) The vast majority of points have no solution, whereas those that do have a unique (x_I, x_R) solution. For this array, the largest number of solutions encountered inside a γ, M_1 window is 22 out of a possible 121. If x_I (or β_I) is fixed, and x_R (or β_R) is allowed to vary, the number of possible solutions ranges from zero to three (Henderson 1964). In other words, with fixed values for γ, M_1, and β_I, there may be as many as three distinct x_R values, each with a solution.

Frequently, a sequence of i-loop (i.e., $\beta_{I,i}$) values, say 1 to 5, possess solutions of a given type, indicating a continuum of this type of solution in the open interval $(\beta_{I,1}, \beta_{I,5})$. At the lower end, the interval is bounded by $\beta_{I,1} - 0.001$, while at the upper end it is bounded by $\beta_{I,6}$. Occasionally, an isolated solution is obtained; this solution actually represents a relatively narrow open β_I interval similar to the foregoing one.

An actual solution is obtained when, during the k loop, an F in the equations in Equation (7.9) has a sign change. The x_R, x_M values are then determined for which F is zero. After this evaluation, the k loop is continued. Upon completion of the k loop, the process is repeated for the next x_I, or β_I, value.

The entropy, S, jump across each shock is evaluated to ensure second law adherence. In the $\gamma = 1$ case, the entropy jump formula is indeterminate. L'Hospital's rule provides

$$\frac{(\Delta S)_i}{R} = \frac{w_i^2 - 1}{2w_i} - \ln w_i, \quad \gamma = 1, \quad i = I, R, M \tag{7.17}$$

where Equation (7.4b) defines w_i and R is the gas constant. In addition, the requirement that the entropy jump across M exceed the sum of the jumps across I and R (Henderson and Menikoff 1998) is evaluated in the form

$$f = \frac{\left(\dfrac{s_2 - s_1}{R}\right) + \left(\dfrac{s_3 - s_2}{R}\right)}{\left(\dfrac{s_4 - s_1}{R}\right)} \tag{7.18a}$$

The condition

$$0 < f < 1 \tag{7.18b}$$

is satisfied, without exception for all cases, including when M is a weak solution shock.

7.4   RESULTS AND DISCUSSION

Parametric results are discussed for $\gamma = 1$, 1.4, and 5/3, where M_1 has its solution onset value followed by 1.4(0.1)2(0.5)6 values. For each γ, M_1 pair, 11 equally spaced β_I values are used in accord with the equations in Equation (7.14). Coverage is densest when $M_1 \leq 2$, and the subsequent presentation favors the $\gamma = 1.4$ cases.

While preparing this material, it became apparent that many additional solutions, besides those provided by the algorithm, were required for proper interpretation. These solutions are for specified values of γ, M_1, and β_I. They support the subsequent discussion but do not alter the algorithm-based tables and figures.

Figures 7.6 and 7.7 depict the solution types with black for (a), red for (b), and blue for (c). Figure 7.6 illustrates the approach when $\gamma = 1.4$ and $M_1 = 5$. (The short, horizontal tabs at each end of a solution bar are a plotting routine artifact. They represent a single algorithm solution.)

In Figure 7.6, where β_I has its minimum value, types (a) and (c) solutions are close to the $\beta_I \rightarrow \beta_{I\mu}$ window edge, where I becomes a Mach wave. When $\beta_I \cong 30°$, there is an (a) $\rightarrow$ (b) transition, M is a normal shock, and $\bar{\theta}_{ss}$ is zero. At the (b) $\rightarrow$ (c)

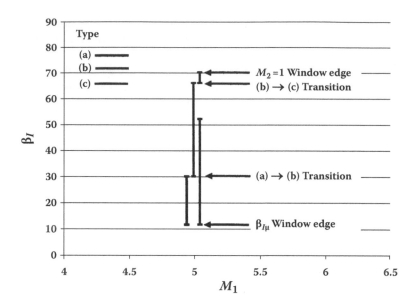

FIGURE 7.6 (See color insert.) M_1, β_I sketch with labeling for $\gamma = 1.4$, $M_1 = 5$.

transition, R is a normal shock, and $\beta_R = 90°$. These normal shocks are smoothly approached from both sides of the transition. As indicated in the figure, the upper window edge is encountered when $\beta_I = \beta_I^*$. In this limit, $M_2 \to 1$ and R becomes a Mach wave.

Types (a) and (b) solutions do not overlap, but both types often overlap (c). When overlap occurs, there are, at least, two distinct solutions with the same γ, M_1, and β_I values. This is evident in Figure 7.5, where types (a) and (c), R inverted, have the same γ, M_1, and β_I values. Although these parameters are the same, the orientation of R, M, and SS are, of course, quite different.

A distinction is made between two solutions of different types with the same γ, M_1, and β_I values, which occur when there is overlap, and two solutions of the same type, also with the same γ, M_1, and β_I values. This latter case is referred to as a double solution. Triple solutions with the same γ, M_1, and β_I values are discussed shortly.

An interesting feature is the two type (c) ranges. The upper segment corresponds to a single point in the algorithm array of solutions. Separate calculations, however, demonstrate a type (c) region within the open interval:

$$\beta_{I,10} < \beta_I < \beta_I^*$$

and there are type (c) solutions in the interval

$$\beta_{I,11} \leq \beta_I < \beta_I^*$$

that are slightly outside the computational solution window. (This aspect would not occur with a sufficiently small ε.)

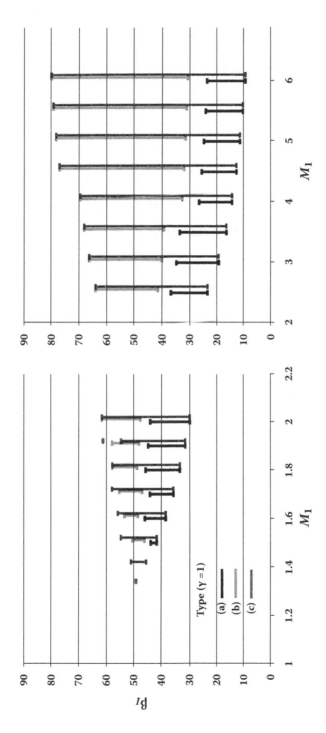

FIGURE 7.7 (See color insert.) Solution types when M_1 equals its onset value, 1.4 (0.1) 2(0.5) 6, (a) $\gamma = 1$.

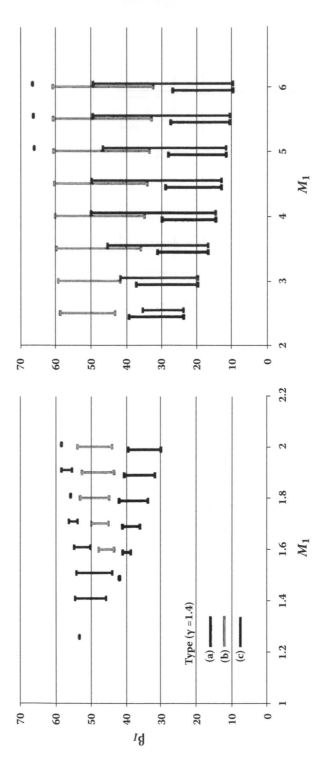

FIGURE 7.7 (*Continued*)   Solution types when M_1 equals its onset value, 1.4 (0.1) 2(0.5) 6, (b) $\gamma = 1.4$.

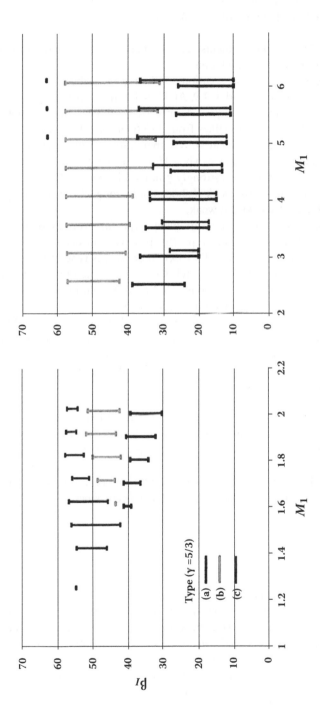

FIGURE 7.7 (Continued)   Solution types when M_1 equals its onset value, 1.4 (0.1) 2(0.5) 6, (c) $\gamma = 5/3$.

Some of the foregoing discussion is specific to Figure 7.6. More general features are now discussed that hold for all γ and M_1 values tested.

1. Only type (c) has split segments. Types (a) and (b) do not have an inverted R shock or overlap with each other.
2. Only type (c) has double solutions. When $\gamma = 1$, these occur between $M_1 = 1.32$ and $M_1 = 1.7$ and at 1.9. At $M_1 = 1.4$, the four (c) algorithm solutions are all doubled. When $\gamma = 1.4$, only $M_1 = 2.5$ has a single double solution. When $\gamma = 5/3$, all M_1 values from 3 to 6 have double solutions. Except for a single doubled solution at $M_1 = 6$, there are multiple doubled solutions per Mach number. For instance, when $M_1 = 5$, the $i = 1, 2, ..., 6$ solutions are doubled. This feature is thus strongly γ and M_1 dependent. For $\gamma = 1$ and 1.4, the double solutions have weak R shocks but strong M shocks. For $\gamma = 5/3$, both the R and M shocks are weak.
3. Triple solutions most often occur when type (a) overlaps doubled type (c). Triple solutions involving type (b), however, do occur. For instance, when $\gamma = 1.4$, $M_1 = 2.662$, and $\beta_I = 39.1°$ there is this type of triple solution (Kalghatgi and Hunt 1975). A search when $\gamma = 1$ failed to reveal any type (b) plus doubled type (c) solutions. This type of solution did occur when $\gamma = 5/3$, $M_1 = 4.5$, and β_I ranged from 33° to 36°.
4. The R and M shocks can be weak or strong, thereby resulting in four different R/M combinations. All four occur, although the category of R strong, M weak is rare. When $\gamma = 1$, no cases occur. When $\gamma = 1.4$, there are two cases: $M_1 = 5.5, 6$ and $i = 11$. When $\gamma = 5/3$, there are two cases: $M_1 = 2$, $i = 10$, and $M_1 = 6$, $i = 11$.
5. There is a type (a) or (c) solution, often both, in the $\beta_I \to \beta_{I\mu}$ limit for all γ and M_1 values tested (Uskov and Chernyshov 2006). The upper window edge, with a type (c) solution, holds for all γ values. In this $M_2 \to 1$ limit, R is a very weak shock and $f \to 1$. For instance, when $\gamma = 1$, $M_1 = 6$, f equals 0.99973.
6. As shown in Uskov and Chernyshov (2006), the type (a) $\to$ (b) transition and type (b) $\to$ (c) transition simultaneously occur when

$$M_1 = \left(\frac{3+\gamma}{2} \right)^{1/2}, \quad \sin\beta_I = \left(\frac{2}{3+\gamma} \right)^{1/2} \tag{7.19}$$

Both M and R are normal shocks. This condition corresponds to I being a Mach wave, where $M_1 \sin\beta_I = 1$, and is on the window's edge. It is excluded from the analysis.

Figure 7.7 contains three β_I, M_1 panels, one each for the three γ values. In this figure and in Figure 7.8, for visual convenience, the horizontal axis has a change in scale at $M_1 = 2$. As indicated, the gap between the top of the type (a) bars and the bottom of the type (b) bars stems from the discrete nature of the algorithm. These bars actually meet where M is a normal shock.

In the $\gamma = 1$ panel, at $M_1 = 1.4$, the type (c) solutions are double solutions, where each pair has slightly different θ_R and θ_M values. The type (c) R shocks are inverted

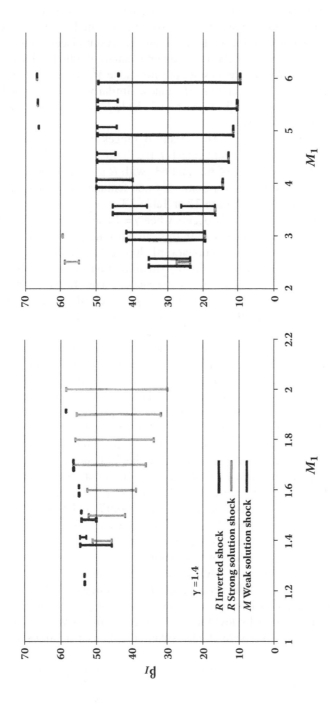

FIGURE 7.8 (See color insert.)   Graph indicating when R is inverted, R is a strong solution shock, and M is a weak solution shock. $\gamma = 1.4$.

with one of the doubled pair having R and M weak, the other strong. This is not a general feature, for instance, at $M_1 = 1.5$, only some of the type (c) solutions are double and not all R shocks are inverted. Note the isolated type (c) solution at $M_1 = 1.90$. This illustrates the second type (c) segment result discussed with respect to Figure 7.6. Because ε was not chosen sufficiently small, the figure does not show that all γ values have isolated type (c) solutions for viable M_1 values. These solutions are evident when $\gamma = 1.4$ and 5/3 at large M_1 values. Similarly, there are type (a) or (c) solutions, often both, when $\beta_I \rightarrow \beta_{I\mu}$ for all γ and viable M_1 values. This is evident at the lower edge of the $\gamma = 1$ panel.

In the $\gamma = 1.4$ panel (e.g., when $M_1 = 3$ and β_I ranges from 42° to 44°) there is type (b) and (c) overlap, which is not evident in the figure. At this M_1 value, there is a (b) $\rightarrow$ (c) transition between β_I of 62° and 63° with R a normal shock. A similar transition occurs at other M_1 values. Some of these features are based on separate solutions and are not evident in Figure 7.7 or Table 7.3.

An unexpected result, which only occurs when $\gamma = 1$ and $3.5 \le M_1 \le 6$, is that $\bar{\beta}_R$, for most type (b) solutions, is negative. In this circumstance, the orientation of the reflected shock is between $\vec{V}_1$ and a downward sloping SS. At the incident shock, β_I has a relatively modest value, and the $\vec{V}_2$ streamline has a sufficiently negative slope that it encounters R at a point below the x-coordinate. In part, this behavior stems from the strong dependence of θ_I and β_R on γ. Overall, the reflected shock has a larger angular variation than the other two shocks. This variation can extend from an inverted position to one below the x-coordinate.

When $\gamma = 1.4$ and $M_1 = 1.5$, the figure shows only type (a) and (c) solutions. Separate calculations, however, provide type (b) solutions in between the $i = 1$ and $i = 2$ algorithm values (i.e., $42.5 \le \beta_I \le 43.5°$). There is thus an (a) $\rightarrow$ (b) transition quickly followed by a (b) $\rightarrow$ (c) transition. This behavior is a consequence of being close to the window edge state given by Equation (7.19).

Figure 7.8 is for $\gamma = 1.4$ and indicates when R is inverted, R is a strong solution shock, and M is a weak solution shock. Note the erratic nature of several features. For instance, the presence of inverted type (c) R shocks is prevalent at large Mach numbers but is not present when $1.8 \le M_1 \le 2$. A strong solution R shock frequently occurs below $M_1 = 2.5$ but less above this Mach number. From $M_1 = 2.5$ to 6, near the $\beta_{I\mu}$ window edge, type (a) has R not inverted and both R and M weak, while type (c) has R inverted and both R and M weak. The lower edge, for this M_1 range, is a mixture of type (a) (R strong) and type (c) (R inverted, M weak), although not in the R strong, M weak combination for a given solution.

Table 7.2 shows conditions at solution onset for the three γ values. These are approximate to the extent that the given onset M_1 value minus 0.01 had no solution. All cases are type (c) and R is inverted, strongly so when $\gamma = 1.4$ and 5/3. In these two cases, $\bar{\beta}_I$ and $\bar{\beta}_R$ differ by less than a degree, and $\bar{\theta}_{SS}$ is very nearly aligned with $\vec{V}_1$. At $\gamma = 1$, there is a type (c) double solution. All onset solutions are close to the $\beta_{I\mu}$ window edge.

Table 7.3 provides results for the 16 algorithm solutions when $\gamma = 1.4$ and $M_1 = 3$. In addition to the type of solution, Mach numbers, and angles, the table shows whether the R and M shocks are weak or strong and if R is inverted. Note that several solutions

TABLE 7.2

Onset of Triple-Point Solutions, Angles in Degrees

γ	M_1	Type	β_I	β_R	β_M	β_{Rd}	β_{Md}	$\bar{\beta}_I$	$\bar{\beta}_R$	$\bar{\beta}_M$	$\bar{\theta}_{ss}$
1	1.32	c	49.31	64.48	64.47	69.70	69.68	130.7	115.5	295.5	351.4
		c	49.31	74.82	74.79	69.70	69.68	130.7	105.1	285.2	351.4
1.4	1.25	c	53.19	54.00	53.98	70.57	70.54	126.8	126.0	306.0	359.5
5/3	1.23	c	54.45	55.26	55.23	70.92	70.88	125.6	124.7	304.8	359.6

TABLE 7.3

Results for $\gamma = 1.4$ and $M_1 = 3$ (n = no, y = yes, wk = weak, str = strong, inv = inverted)

Solution	Type	M_2	M_3	M_4	$\bar{\beta}_I$	$\bar{\beta}_R$	$\bar{\beta}_M$	$\bar{\theta}_{ss}$	R inv	R	M
1	a	2.996	0.8035	0.7957	160.5	70.30	250.7	33.02	n	str	str
2		2.700	1.057	0.5932	156.1	55.23	259.1	25.34	n	wk	str
3		2.452	1.199	0.5148	151.6	44.76	263.9	16.23	n	wk	str
4		2.228	1.250	0.4837	147.2	37.61	267.2	7.841	n	wk	str
5		2.020	1.244	0.4752	142.8	32.80	269.9	0.2241	n	wk	str
6	b	1.824	1.202	0.4814	138.4	29.91	272.4	353.3	n	wk	str
7		1.639	1.141	0.5004	134.0	28.84	274.8	346.8	n	wk	str
8		1.465	1.071	0.5342	129.5	29.82	277.5	340.7	n	wk	str
9		1.301	1.002	0.5903	125.1	33.53	280.7	334.9	n	wk	str
10		1.146	0.9509	0.6927	120.7	41.84	285.4	329.5	n	str	str
11	c	2.996	1.193	1.188	160.5	122.0	301.9	327.3	y	wk	wk
12		2.700	1.374	1.080	156.1	122.4	298.7	326.4	y	wk	wk
13		2.452	1.478	1.064	151.6	121.6	298.2	326.3	y	wk	wk
14		2.228	1.523	1.105	147.2	120.0	299.5	326.5	y	wk	wk
15		2.020	1.539	1.193	142.8	118.0	302.1	327.3	y	wk	wk
16		1.824	1.560	1.354	138.4	116.3	306.6	329.3	y	wk	wk

are subsonic in region 3, while other solutions have supersonic flow in region 4. Also note that M_3 and M_4 may simultaneously be subsonic or supersonic. R is inverted only for type (c) shocks, where both R and M shocks are weak. Types (a) and (b) solutions have weak R shocks and strong M shocks, except for the first solution. Between solutions 5 and 6, there is an (a) $\rightarrow$ (b) transition with a normal M shock. This is apparent from $\bar{\theta}_{ss}$, which goes through zero between these two solutions. As indicated earlier, there is a split type (c) solution when $\bar{\beta}_I = 117°$. Hence, there is also a (b) $\rightarrow$ (c) transition with a normal R shock. There are no double type (c) solutions, but there are solutions where (a) and (c) overlap and where (b) and (c) overlap.

TABLE 7.4

Two Triple-Solution Cases When $\gamma = 1.4$ and $M_1 = 2.5$ (n = no, y = yes, wk = weak, str = strong, inv = inverted)

Type	β_I	$\bar{\beta}_I$	$\bar{\beta}_R$	$\bar{\beta}_M$	$\bar{\theta}_{ss}$	M_2	M_3	M_4	R inv	R	M
a	23.64	156.4	71.76	252.1	279.4	2.497	0.7600	0.7550	n	str	str
c	23.64	156.4	156.0	336.1	359.5	2.497	2.479	2.479	y	wk	wk
c	23.64	156.4	126.2	306.2	333.1	2.497	1.271	1.269	y	wk	wk
a	34.32	144.7	46.56	266.6	7.849	1.939	1.033	0.5235	n	wk	str
c	34.32	144.7	134.8	323.8	345.7	1.939	1.909	1.902	y	wk	wk
c	34.32	144.7	124.1	309.7	335.0	1.939	1.525	1.384	y	wk	wk

The type (a) $\rightarrow$ (b) $\rightarrow$ (c) sequence in Table 7.3 represents a clockwise rotation of $\bar{\theta}_{ss}$ from above the x-axis to below it at the start of type (b). Partway through type (c), $\bar{\theta}_{ss}$ reverses its rotational direction. This reversal does not include the split (c) segment.

Table 7.4 has data for two triple solutions that occur when $\gamma = 1.4$ and $M_1 = 2.5$. When $\gamma = 5/3$, there are many triple solutions, since, when $i = 1$, type (c) is doubled for all $M_1 \geq 3$. As in Table 7.2, the first type (c) $\beta_I = 23.64°$ solution has the I and R shocks nearly coincident and $\bar{\theta}_{ss}$ nearly horizontal.

Our results, when available, agree with those in Uskov and Chernyshov (2006), which are for special cases, such as when M or R is a normal shock, or an extreme triple-point configuration, such as occur near the window's edge. Their results are limited to $\gamma = 1.4$. Nevertheless, this comparison makes evident the need for a smaller ε value and a larger N value than utilized in the present study.

8 Derivatives When the Upstream Flow Is Nonuniform

8.1 PRELIMINARY REMARKS

The uniform upstream flow constraint of Chapter 4 is removed. Of course, if the shock is two-dimensional (axisymmetric), the nonuniform upstream flow must also be two-dimensional (axisymmetric). This constraint does not alter the jump conditions but substantially alters the tangential and normal derivatives. Earlier assumptions are retained, such as a steady flow of a perfect gas. The flow plane approach means the jump conditions hold in a three-dimensional flow, as before. The normal derivatives still require a two-dimensional or axisymmetric flow. The homenergetic aspect is retained (except in Section 8.6), but the nonuniform upstream flow may be rotational or irrotational. In Section 8.7, an illustrative model is provided where the nonuniform upstream flow is irrotational.

A number of notational changes are required. Angles, such as β, when measured from the x-coordinate, see Figure 8.1, are primed, as compared to unprimed angles that are measured from $\bar{V}_1$. In the preceding analysis, $\bar{V}_1$ is aligned with the x-coordinate, and there is no difference between primed and unprimed angles. The angles that $\bar{V}_1$ and $\bar{V}_2$ have, relative to the x-coordinate, are denoted as δ'_1 and δ'_2, respectively. The various angles satisfy

$$\theta = \delta'_2 - \delta'_1 = \theta' - \delta'_1, \quad \theta' - \delta'_2, \quad \beta = \beta' - \delta'_1, \quad \beta - \theta = \beta' - \theta' \tag{8.1}$$

The downstream streamline, or $\bar{V}_2$, angle is denoted as θ' (see Figure 8.1). The $d\beta/ds$ derivative was denoted as β'; it now is written as β_s. Tangential shock and normal derivatives are denoted with s and n subscripts, respectively.

The $-\beta_s$ parameter is now the local shock curvature in the flow plane; the actual (geometric) curvature, $-\beta'_s$, requires the shock angle be measured from the x-coordinate. This curvature equals $-\beta_s + d\delta'_1/ds$. Similarly, when the flow is axisymmetric, the local transverse curvature of the shock, in a plane normal to the shock and flow plane, is now $\cos\beta/y$, not $\cos\beta'/y$.

The convenient Appendix E notation is retained, where w is still the square of the normal component of the Mach number, $m\sin^2\beta$. Note that X, Y, and Z, without a subscript, use w. On the other hand, an isentropic relation, such as

$$p_o = p_1 \, X_1^{\gamma/(\gamma-1)} \tag{8.2a}$$

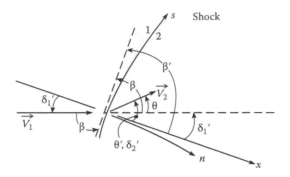

FIGURE 8.1   Shock schematic in the flow plane.

utilizes

$$X_1 = 1 + \frac{\gamma - 1}{2} m = 1 + \frac{\gamma - 1}{2} M_1^2 \tag{8.2b}$$

Throughout the subsequent analysis, $\bar{p}_1$, $\bar{\rho}_1$, and $\bar{V}_1$ normalize the pressure, density, and velocity or velocity components, respectively, where an overbar denotes a dimensional variable or parameter. Lengths are normalized with a constant reference length, $\bar{R}$, which could be the radius of a sphere or cylinder. We thus define

$$s = \frac{\bar{s}}{\bar{R}}, \quad n = \frac{\bar{n}}{\bar{R}}, \quad \tilde{s} = \frac{\bar{\tilde{s}}}{\bar{R}}, \quad \tilde{n} = \frac{\bar{\tilde{n}}}{\bar{R}}, \quad x = \frac{\bar{x}}{\bar{R}}, \quad y = \frac{\bar{y}}{\bar{R}} \tag{8.3}$$

where $\tilde{s}$ and $\tilde{n}$ are intrinsic coordinates. Nondimensional variables and derivatives are written as

$$u_2 = \frac{\bar{u}_2}{\bar{V}_1}, \quad p_2 = \frac{\bar{p}_2}{\bar{p}_1}, \ \cdots \tag{8.4a}$$

$$M_{1s} = \left(\frac{\partial M}{\partial s}\right)_1, \quad V_{1s} = \frac{1}{\bar{V}_1}\left(\frac{\partial \bar{V}}{\partial s}\right)_1, \quad p_{1s} = \frac{1}{\bar{p}_1}\left(\frac{\partial \bar{p}}{\partial s}\right)_1, \quad p_{2n} = \frac{1}{\bar{p}_1}\left(\frac{\partial \bar{p}}{\partial n}\right)_2, \ \cdots \tag{8.4b}$$

This normalization does not imply that $\bar{p}_1$, $\bar{\rho}_1$, or $\bar{V}_1$ are constants. As a result of the normalization, the inverse of the parameter

$$\gamma m = \gamma M_1^2 = \frac{\gamma \bar{V}_1^2}{\gamma \bar{p}_1 / \bar{\rho}_1} = \frac{\bar{\rho}_1 \bar{V}_1^2}{\bar{p}_1} \tag{8.5}$$

is frequently encountered. In this chapter, generally only nondimensional parameters and variables are used.

8.2 JUMP CONDITIONS

Because of the complexity of the normal derivatives, given later, a vector/matrix formulation is introduced. In vector notation, define

$$q_j = (u, v, p, \rho)_2, \qquad j = 1, ..., 4 \tag{8.6}$$

where, for example, $q_2 = v_2$. The jump conditions are

$$q_j = f_j \tag{8.7a}$$

or

$$\vec{q} = \vec{f} \tag{8.7b}$$

where the f_j are given in Appendix F.1. Of course, these stem from the Appendix E.1 equations.

8.3 TANGENTIAL DERIVATIVES

Again, in vector notation, derivatives, in the flow plane, along the upstream side of the shock are written as

$$\chi_i = (M_{1s}/M_1, V_{1s}, p_{1s}, \rho_{1s}, \beta_s), \qquad i = 1, ..., 5 \tag{8.8}$$

and the desired derivatives as

$$q_{js} = \frac{\partial q_j}{\partial s} = \left(u_{2s}, v_{2s}, p_{2s}, \rho_{2s} \right), \qquad j = 1, \cdots, 4 \tag{8.9}$$

Equation (8.7a) is differentiated with respect to s, with the result

$$q_{js} = \sum_{i=1}^{5} g_{ji}\chi_i \tag{8.10a}$$

or

$$\vec{q}_s = g\vec{\chi} \tag{8.10b}$$

where g is a 4×5 matrix. The g_{ji} elements are summarized in Appendix F.2. They are functions only of γ, β, and M_1.

Note that there is no direct dependence on δ_1'. The dependence on $d\delta_1'/ds$, however, is accounted for in β_s'. In Section 8.7, a flow is analyzed where $\beta_s' = 0$, but β_s is variable. This situation occurs because the shock is straight (i.e., $\beta_s' = 0$), but there is local curvature, because $d\delta_1'/ds$ is varying.

To illustrate the derivation, v_{2s} is obtained:

$$\bar{v}_2 = \frac{2}{\gamma+1}\bar{V}_1 \frac{X}{m\sin\beta} \tag{8.11}$$

$$\left(\frac{\partial\bar{v}}{\partial s}\right)_2 = \frac{2}{\gamma+1}\left(\frac{X}{m\sin\beta}\frac{\partial\bar{V}_1}{\partial s} + \frac{\gamma-1}{2}\frac{\bar{V}_1}{m\sin\beta}\frac{dw}{ds} - \frac{2\bar{V}_1 X}{m^2\sin\beta}M_1 M_{1s} - \frac{\bar{V}_1 X}{m\sin^2\beta}\beta_s\cos\beta\right)$$

$$v_{2s} = \frac{2}{\gamma+1}\left(\frac{X}{m\sin\beta}V_{1s} + \frac{\gamma-1}{2}\frac{1}{m\sin\beta}\frac{dw}{ds} - \frac{2X}{m\sin\beta}\frac{M_{1s}}{M_1} - \frac{X}{w}\beta_s\cos\beta\right) \tag{8.12a}$$

The *dw/ds* derivative is

$$\frac{dw}{ds} = 2M_1\frac{dM_1}{ds}\sin^2\beta + 2M_1^2\sin\beta\cos\beta\frac{d\beta}{ds} = 2w\frac{M_{1s}}{M_1} + \frac{4}{\gamma+1}XA\beta_s \tag{8.13}$$

Elimination of *dw/ds* from Equation (8.12a), after simplification, yields

$$v_{2s} = -\frac{4}{\gamma+1}\frac{1}{m\sin\beta}\frac{M_{1s}}{M_1} + \frac{2}{\gamma+1}\frac{X}{m\sin\beta}V_{1s} - \frac{2}{\gamma-1}\frac{1-\frac{\gamma-1}{2}w}{w}\beta_s\cos\beta \tag{8.12b}$$

which is in accord with the listed g_{2i}.

The first four χ_i are not independent. In view of Equation (8.5), we have

$$\frac{M_{1s}}{M_1} = V_{1s} - \frac{1}{2}P_{1s} + \frac{1}{2}\rho_{1s} \tag{8.14}$$

It is convenient, however, to retain the V_{1s} and M_{1s} terms and not replace one with the other. When the upstream flow is uniform, the first four χ_i are zero and Equation (8.10a) agrees with Appendix E.2.

This section concludes by obtaining the derivative of θ starting with its equation in Appendix E.1:

$$\tan\theta = \frac{1}{\tan\beta}\frac{Z}{\frac{\gamma+1}{2}m - Z} \tag{8.15}$$

$$\ln\tan\theta = -\ln\tan\beta + \ln Z - \ln\left(\frac{\gamma+1}{2}M_1^2 - Z\right)$$

$$\frac{\theta_s}{\tan\theta\cos^2\theta} = -\frac{\beta_s}{\tan\beta\cos^2\beta} + \frac{1}{Z}\frac{dw}{ds} - \frac{(\gamma+1)M_1 M_{1s} - (dw/ds)}{\frac{\gamma+1}{2}m - Z}$$

$$
\frac{\theta_s}{\sin\theta\,\cos\theta} = -\frac{\beta_s}{\sin\beta\,\cos\beta} + \left(\frac{1}{Z} + \frac{1}{\frac{\gamma+1}{2}m - Z}\right)\left(2w\frac{M_{1s}}{M_1} + \frac{4}{\gamma+1}XA\beta_s\right)
$$

$$
-\frac{(\gamma+1)m}{\frac{\gamma+1}{2}m - Z}\frac{M_{1s}}{M_1}
$$

$$
\frac{\theta_s}{\sin\theta\,\cos\theta} = -\frac{\beta_s}{\sin\beta\,\cos\beta} + \frac{(\gamma+1)m}{Z\left(\frac{\gamma+1}{2}m - Z\right)}\frac{M_{1s}}{M_1} + \frac{2mXA}{Z\left(\frac{\gamma+1}{2}m - Z\right)}\beta_s \qquad (8.16)
$$

One can show that

$$
\sin\theta\,\cos\theta = \frac{Z\left(\frac{\gamma+1}{2}m - Z\right)}{X^2 B}\sin\beta\,\cos\beta \qquad (8.17)
$$

where $X^2 B$ is given by Equation (5.13c). Equation (8.16) now becomes

$$
\theta_s = \frac{2A}{XB}\frac{M_{1s}}{M_1} + \frac{\left[\frac{\gamma+1}{2}m\,(1+w) + 1 - 2w - \gamma w^2\right]}{X^2 B}\beta_s \qquad (8.18)
$$

When M_{1s} is zero (uniform freestream); θ_s is zero when β_s is zero, a trivial result, or when the term in the square bracket is zero. The bracket term is zero when β has its uniform upstream flow detachment value, β_d, given by Equation (7.10). Detachment, with a uniform freestream, occurs when θ has a maximum value relative to β—that is, $(\partial\theta/\partial\beta) = 0$, which, in turn, requires $\theta_s = 0$. As indicated by Equation (8.18), detachment (i.e., $\theta_s = 0$) is more involved when the freestream is not uniform. When $M_{1s} = 0$, θ_s agrees with Chapter 5 results, such as the $(\partial\theta/\partial s)_2$ equation below Equation (5.3), for a normal shock.

8.4   NORMAL DERIVATIVES

As discussed in Section 4.4, the angle θ, between ξ_1 and the x-coordinate becomes β when the coordinates used with the Euler equations are adjusted to a shock. When the upstream flow is uniform, and $\vec{V}_1$ is aligned with x, it is convenient, because of the jump conditions, to measure β relative to $\vec{V}_1$. Now, however, β becomes β', which is measured relative to the x-coordinate. From Equation (8.1), we have the connection

$$
\beta' = \beta + \delta_1', \quad \beta_s' = \beta_s + \delta_{1s}' \qquad (8.19)
$$

The two vectors are

$$q_n = (u_{2n}, v_{2n}, p_{2n}, \rho_{2n}) \tag{8.20}$$

$$\chi_i' = \left(u_{2s}, v_{2s}, p_{2s}, \rho_{2s}, \beta_s', \frac{\sigma}{y}\alpha_1 \right) \tag{8.21}$$

where

$$\alpha_1 = u_2 \sin\beta' - v_2 \cos\beta' \tag{8.22a}$$

The $\sigma\alpha_1/y$ term stems from the axisymmetric term in the continuity equation. The normal derivative equations have the form

$$q_{jn} = \sum_{i=1}^{6} h_{ji} \, \chi_i' \tag{8.23a}$$

or

$$\vec{q}_n = h \, \vec{\chi}' \tag{8.23b}$$

where h is a 4×6 matrix. The rightmost two elements in χ' depend on the longitudinal curvature, $-\beta_s'$, and on the transverse curvature that occurs in an axisymmetric flow. Along with γ and M_1, the h_{ji} will depend on q (i.e., u_2, v_2, p_2, and ρ_2). No attempt has been made to eliminate u_2, v_2, ..., and u_{2s}, v_{2s}, ..., in favor of M_{1s}, ..., from the right side of Equation (8.23a). (This could be done using symbolic manipulation software.)

Initially, the sequential procedure for obtaining the normal derivatives, discussed after Equations (4.11), is followed. From the tangential momentum equation in Equation (4.10), u_{2n} is

$$u_{2n} = -\frac{u_2}{v_2} u_{2s} - \frac{1}{\gamma m \rho_2 v_2} p_{2s} - u_2 \beta_s' \tag{8.24a}$$

Continuity can be written as

$$\rho_2 v_{2n} + v_2 \rho_{2n} = \alpha_2 \tag{8.25a}$$

where

$$\alpha_2 = -\left(\rho_2 u_{2s} + u_2 \rho_{2s} + \rho_2 v_2 \beta_s' + \frac{\sigma}{y}\rho_2 \alpha_1 \right) \tag{8.22b}$$

By eliminating p_{2n}/ρ_2 from the normal momentum equation and the normal derivative of the homenergetic equation, a second equation with v_{2n} and ρ_{2n} as the unknowns is obtained:

$$v_2 v_{2n} + \frac{p_2}{m\rho_2^2} \rho_{2n} = \alpha_3 \tag{8.25b}$$

where

$$\alpha_3 = (\gamma - 1) u_2 u_{2n} - \gamma u_2 v_{2s} + \gamma u_2^2 \beta_s' \qquad (8.22c)$$

The equations in Equation (8.25) are solved, with the result

$$\Delta v_{2n} = \frac{1}{m} \frac{p_2}{\rho_2^2} \alpha_2 - v_2 \alpha_3 \qquad (8.26a)$$

$$\Delta \rho_{2n} = \rho_2 \alpha_3 - v_2 \alpha_2 \qquad (8.26b)$$

where the left-side determinant is

$$\Delta = \frac{p_2}{m \rho_2} - v_2^2 \qquad (8.27)$$

This determinant is zero only when the shock becomes a Mach wave (see Equation 9.80k). Finally, the pressure derivative is given by

$$p_{2n} = \frac{p_2}{\rho_2} \rho_{2n} - (\gamma - 1) m \rho_2 (u_2 u_{2n} + v_2 v_{2n}) \qquad (8.26c)$$

As mentioned, with

$$M_{1s} = V_{1s} = p_{1s} = \rho_{1s} = 0 \qquad (8.28)$$

the tangential derivatives check against Appendix F.2. With $\beta' = \beta$ and Equation (8.28), u_{2n}, given by Equation (8.24a), reduces to the result given in Appendix E.3, including the g_1 equation. A much more tedious, but successful, check for v_{2n} yielded the result in Appendix E.3, including the g_2 equation. This type of check should be done with symbolic manipulation software.

The α_1, α_2, and Δ parameters do not depend on any normal derivative, whereas α_3 depends on u_{2n}. Elimination of u_{2n} from α_3 yields

$$\alpha_3 = -(\gamma - 1) \frac{u_2^2}{v_2} u_{2s} - \gamma u_2 v_{2s} - \frac{\gamma - 1}{\gamma} \frac{u_2}{m \rho_2 v_2} p_{2s} + u_2^2 \beta_s' \qquad (8.29)$$

By substituting this relation into Equation (8.26a,b), all normal derivatives are removed from the right side of the v_{2n} and ρ_{2n} equations, with the result

$$\Delta v_{2n} = \left[-\frac{p_2}{m \rho_2} + (\gamma - 1) u_2^2 \right] u_{2s} + \gamma u_2 v_2 v_{2s} + (\gamma - 1) \frac{u_2}{\gamma m \rho_2} p_{2s}$$

$$- \frac{p_2 u_2}{m \rho_2^2} \rho_{2s} - v_2 \left(\frac{p_2}{m \rho_2} + u_2^2 \right) \beta_s' - \frac{p_2}{m \rho_2} \frac{\sigma}{y} \alpha_1 \qquad (8.24b)$$

$$\Delta \rho_{2n} = \frac{\rho_2}{v_2}\left[-(\gamma-1)u_2^2 + v_2^2\right]u_{2s} - \gamma\rho_2\,u_2\,v_{2s} - (\gamma-1)\frac{u_2}{\gamma m v_2}\,p_{2s}$$

$$+ u_2\,v_2\,\rho_{2s} + \rho_2\left(u_2^2 + v_2^2\right)\beta_s' + \rho_2 v_2\frac{\sigma}{y}\alpha_1 \qquad (8.24c)$$

By replacing the ρ_{2n}, u_{2n}, and v_{2n} derivatives in Equation (8.26c), we finally obtain

$$\Delta p_{2n} = \gamma m \rho_2 v_2\left[\frac{p_2}{m\rho_2} - (\gamma-1)u_2^2\right]u_{2s} - \gamma m \rho_2\,u_2\left[\frac{p_2}{m\rho_2} + (\gamma-1)v_2^2\right]v_{2s}$$

$$-(\gamma-1)u_2v_2p_{2s} + \frac{\gamma p_2 u_2 v_2}{\rho_2}p_{2s} + \gamma p_2\left(u_2^2 + v_2^2\right)\beta_s' + \gamma p_2 v_2\frac{\sigma}{y}\alpha_1 \qquad (8.24d)$$

The equations in Equation (8.24) are the desired normal derivatives. This result is summarized in Appendix F.3.

8.5   INTRINSIC COORDINATE DERIVATIVES

One motivation for the study in this section is to verify the curved shock theory (CST) of Molder (2012). (This verification has been accomplished.) This theory utilizes the standard, two-dimensional or axisymmetric, steady flow gas dynamic assumptions, except the upstream flow may be nonuniform. In CST, the focus is on two parameters: these are the streamline derivatives, just downstream of the shock, of the pressure and the streamline's inclination angle δ_2'. These two parameters are functions of the two shock wave curvatures and the upstream values of the two streamline derivatives and the vorticity. Although there is some overlap in results, the CST approach and that used here are quite different. The subsequent analysis provides equations for the CST parameters, $(\partial p/\partial \tilde{s})_2$ and $(\partial \delta_2'/\partial \tilde{s})$, when the freestream is nonuniform.

The streamline derivative is given by Equation (5.10a), where we note that

$$\beta - \theta = \beta' - \theta' = \beta' - \delta_2' \qquad (8.30)$$

We thus obtain, for the pressure,

$$\left(\frac{\partial p}{\partial \tilde{s}}\right)_2 = \frac{1}{B^{1/2}}(A\,p_{2s} + p_{2n}) \qquad (8.31)$$

The δ_2' streamline derivative cannot be obtained from Equation (8.18). It is conveniently obtained from the streamline's transverse momentum equation, Equation (5.22), when written with intrinsic coordinates:

$$\frac{\partial \theta'}{\partial \tilde{s}} + \frac{1}{\gamma m \rho V^2}\frac{\partial p}{\partial \tilde{n}} = 0 \qquad (8.32)$$

where, at the shock, $\theta' = \delta_2'$. The $\tilde{n}$ derivative is given by Equation (5.10b) with the result

$$\left(\frac{\partial p}{\partial \tilde{n}}\right)_2 = \frac{1}{B^{1/2}}(p_{2s} - A p_{2n}) \tag{8.33}$$

Hence, the desired δ_2' derivative is

$$\frac{\partial \delta_2'}{\partial \tilde{s}} = \frac{1}{\gamma m \rho_2 V_2^2 B^{1/2}}(-p_{2s} + A p_{2n}) \tag{8.34}$$

where $V_2^2 = u_2^2 + v_2^2$. Both CST parameters linearly depend on the p_{2s} and p_{2n} derivatives.

In CST, the p and δ_2' streamline derivatives, at state 2, explicitly depend on the upstream vorticity. This is not the case here, because the treatment automatically includes the presence of any upstream vorticity, which is the subject of the next section.

8.6 VORTICITY

The vorticity, just downstream of a curved shock, is provided by Equation (6.6) when the upstream flow is uniform. Its nonuniform upstream flow counterpart is derived, but without the homenergetic requirement. The derivation is straightforward without this requirement because the shock is two-dimensional or axisymmetric. This section is largely based on Emanuel (2011).

We start with Crocco's equation minus the unsteady term. With the aid of Equation (6.1), we obtain the dimensional result

$$\omega = -\frac{T}{v}\frac{\partial S}{\partial s} + \frac{1}{v}\frac{\partial h_o}{\partial s} \tag{8.35}$$

or

$$\frac{dS}{ds} = -\frac{v}{T}\omega + \frac{1}{T}\frac{dh_o}{ds} \tag{8.36}$$

where partial derivatives become ordinary derivatives along the shock.

The entropy change across a shock is

$$S_2 - S_1 = -\frac{R}{\gamma-1} \ln\left[\left(\frac{\gamma+1}{2}\right)^{(\gamma+1)} \frac{w^\gamma}{X^\gamma Y}\right] \tag{8.37}$$

where R is the gas constant. The derivative with respect to w results in

$$\frac{dS_2}{dw} = \frac{dS_1}{dw} + \frac{\gamma R}{2}\frac{Z^2}{wXY} \tag{8.38a}$$

or

$$\frac{dS_2}{ds} = \frac{dS_1}{ds} + \frac{\gamma R}{2}\frac{Z^2}{wXY}\frac{dw}{ds}$$

(8.38b)

The entropy derivatives are replaced with Equation (8.36), to obtain

$$\omega_2 = \frac{v_1}{v_2}\frac{T_2}{T_1}\omega_1 - \frac{\gamma R}{2}\frac{T_2}{v_2}\frac{Z^2}{wXY}\frac{dw}{ds} + \frac{1}{v_2}\left[\left(\frac{dh_o}{ds}\right)_2 - \frac{T_2}{T_1}\left(\frac{dh_o}{ds}\right)_1\right]$$

(8.39a)

The following relations

$$\frac{v_1}{V_1} = \sin\beta, \quad \frac{v_2}{V_1} = \frac{2}{\gamma+1}\frac{X}{M_1^2\sin\beta}$$

(8.40a,b)

$$\frac{T_2}{T_1} = \left(\frac{2}{\gamma+1}\right)^2\frac{XY}{w}, \quad \frac{p_2}{p_1} = \frac{2}{\gamma+1}Y, \quad \left(\frac{dh_o}{ds}\right)_2 = \left(\frac{dh_o}{ds}\right)_1$$

(8.40c,d,e)

are introduced into Equation (8.39a), with the result

$$\omega_2 = \frac{p_2}{p_1}\omega_1 - \frac{1}{\gamma+1}V_1\frac{Z^2}{w^2X}\sin\beta\frac{dw}{ds} - \frac{\gamma-1}{\gamma+1}\frac{(1+\gamma w)Z}{V_1 X\sin\beta}\left(\frac{dh_o}{ds}\right)_1$$

(8.39b)

Equations (8.40b,c,d) are the jump conditions for v, T, and p. Equation (8.40e) stems from h_o being a constant across a shock. By replacing dw/ds with Equation (8.13), the final nondimensional result is obtained:

$$\omega_2 = \frac{p_2}{p_1}\omega_1 - \frac{\gamma-1}{\gamma+1}\frac{(1+\gamma w)Z}{V_1^2 X\sin\beta}\left(\frac{dh_0}{ds}\right)_1 - \left(\frac{2}{\gamma+1}\frac{Z^2}{wX}\right)\frac{\sin\beta}{M_1}\left(\frac{dM}{ds}\right)_1$$
$$- \left(\frac{2}{\gamma+1}\frac{Z^2}{wX}\right)\beta_s\cos\beta$$

(8.39c)

where the vorticities are normalized with V_1 and a convenient length scale. Note that ω_2 only depends on local parameters—that is, V_1, M_1, β, β_s, $(dM/ds)_1$, $(dh_o/ds)_1$, and ω_1. As written, there is no dependence on the $\cos\beta/y$ curvature, and the β angle is unprimed. If β_s, however, is replaced with Equation (8.19), the transverse curvature can be introduced through the $(\partial\delta_1'/\partial s)$ term. This derivative is first written in terms of intrinsic coordinate derivatives. The continuity equation, in intrinsic coordinates, then involves both the $\partial\delta_1'/\partial\tilde{n}$ derivative and the $\cos\beta/y$ curvature when the flow is axisymmetric.

The p_2/p_1 coefficient stems from continuity, $(\rho v)_2 = (\rho v)_1$, and can considerably amplify ω_1. The $(dh_o/ds)_1$ and $(dM/ds)_1$ terms account for upstream gradients not associated with vorticity. For instance, if the upstream flow is a supersonic, cylindrical source flow, as in the next section, ω_1 and $(dh_o/ds)_1$ are zero but $(dM/ds)_1$ is not zero when the shock is not cylindrical. The β_s term in Equation (8.39c) agrees with Equation (6.6) and represents the local longitudinal curvature contribution. Note that Equation (8.39c) appreciably differs from

$$\omega_2 = \omega_1 - \frac{2}{\gamma+1} V_1 \frac{Z^2}{wX} \beta_s \cos\beta \tag{8.41}$$

that sometimes is incorrectly used.

8.7 SOURCE FLOW MODEL

The model consists of a supersonic, cylindrical source flow that impinges on a planar, straight shock. Only a wedge slice of the source flow is relevant. The upstream flow is cylindrically symmetric whose solution only depends on the radial distance from the (virtual) source. (With characteristic theory, it is possible to design an asymmetric nozzle with a supersonic, wedge source flow at its exit.) Flow conditions just upstream of the straight shock are therefore nonuniform, homenergetic, and irrotational. Nevertheless, the flow downstream of the shock is rotational (see Problem 22). (There is an entropy gradient along the downstream side of the shock.) There is no presumption that the model represents an actual flow.

A similar analysis, by the author, utilizes a point source that generates a spherically symmetric, supersonic flow that impinges on a conical shock whose symmetry axis passes through the point source. This is a relatively simple, analytical, test model for an axisymmetric flow. It has many of the same features as the two-dimensional test model.

A Cartesian coordinate system (Figure 8.2) is utilized whose origin is the center of the cylindrical source flow. The shock wave angle, β', is a constant, whereas the other angles vary. The solution is constrained to a φ' range of zero to φ'^*, where the shock becomes a Mach wave. This occurs despite M_1 increasing with φ', because the normal Mach number component, M_{1n}, decreases.

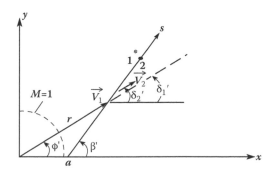

FIGURE 8.2 Flow schematic.

The nondimensional solution for the flow just upstream and downstream of the shock only depends on the ratio of specific heats, γ, the fixed angle β', the normal component of the state a Mach number, M_{1na}, and φ'. (State a is where the shock and the x-coordinate intersect, see Figure 8.2.) For a given configuration, only φ' is variable, and the β' angle is (arbitrarily) chosen such that the shock is a weak solution shock at state a. A wedge (not shown in the figure) for generating the shock, with a curved surface and a sharp tip at state a would have a positive inclination, δ'_{2a}, at the tip.

The strength of the source flow, per unit depth, $\dot{m}$, is provided by continuity

$$\dot{m} = 2\pi \bar{r} \rho V \tag{8.42}$$

where an overbar represents a dimensional length parameter, and, at the shock,

$$V_1 = a_1 M_1 = a_o \frac{M_1}{X_1^{1/2}} \tag{8.43}$$

At an arbitrary point on the shock, Equation (8.42) reduces to

$$r = \frac{\bar{r}}{\bar{R}} = \frac{1}{M_1} X_1^{(\gamma+1)/[2(\gamma-1)]} \tag{8.44}$$

where all lengths are normalized with

$$\bar{R} = \frac{1}{\rho_o a_o} \frac{\dot{m}}{2\pi} \tag{8.45}$$

When $M_1 = 1$, there is a sonic circle with a radius

$$r^* = \left(\frac{\gamma+1}{2} \right)^{(\gamma+1)/[2(\gamma-1)]} = 1.728 \tag{8.46}$$

when $\gamma = 1.4$. This circle is sketched in Figure 8.2. The derivative of Equation (8.44) yields

$$\frac{dr}{dM_1} = \frac{M_1^2 - 1}{M_1^2} X_1^{(3-\gamma)/[2(\gamma-1)]} \tag{8.47}$$

which indicates an extremum when $M_1 = 1$. Figure 8.3 is a sketch of M_1 versus r, which shows that r has a minimum at r^*. Only the supersonic upper branch is of interest.

The normal component of the Mach number is

$$M_{1n} = M_1 \sin (\beta' - \varphi') \tag{8.48}$$

Hence, at state a, where $\varphi' = 0$, we have

$$M_{1a} = \frac{M_{1na}}{\sin\beta'} \tag{8.49}$$

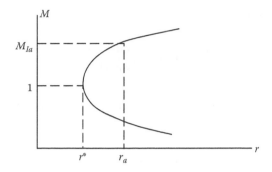

FIGURE 8.3   Cylindrical flow sketch for M_1 vs. r.

where M_{1na} and β' are prescribed, and M_{1na} should be well in excess of unity. This is necessary if φ'^* is to exceed, say, $40°$. We now have, at state a,

$$r_a = \frac{1}{M_{1a}} X_{1a}^{(\gamma+1)/[2(\gamma-1)]} \tag{8.50}$$

$$\tan\delta'_{2a} = \frac{1}{\tan\beta'} \frac{Z_{1a}}{\frac{\gamma+1}{2} M_{1a}^2 - Z_{1a}} \tag{8.51}$$

From the law of sines, the radial coordinate at the shock is

$$r = \frac{r_a \sin\beta'}{\sin(\beta' - \varphi')} \tag{8.52}$$

Given γ, β', φ', and M_{1na}, this equation links r and φ'. Equation (8.44) then connects M_1 and r. In the following analysis, it is convenient to view M_1, instead of φ', as the prescribed independent parameter, although computational results revert to φ'. Equation (8.44) then provides r, while φ' is given by the inversion of Equation (8.52):

$$\varphi' = \beta' - \sin^{-1}\left(\frac{r_a \sin\beta'}{r}\right) \tag{8.53}$$

The derivate of φ', needed shortly, is

$$\frac{d\varphi'}{dr} = \frac{1}{r\left[\left(\frac{r}{r_a}\sin\beta'\right)^2 - 1\right]^{1/2}} \tag{8.54a}$$

or, more conveniently,

$$\frac{dr}{d\varphi'} = \left(r_a \sin\beta'\right)\frac{\cos\left(\beta' - \varphi'\right)}{\sin^2\left(\beta' - \varphi'\right)} \tag{8.54b}$$

The length along the shock, measured from state a, is also given by the law of sines:

$$s = \frac{r_a \sin\varphi'}{\sin\left(\beta' - \varphi'\right)} \tag{8.55}$$

Its derivative is

$$\frac{ds}{d\varphi'} = \frac{r^2}{r_a \sin\beta'} \tag{8.56}$$

which requires the use of Equation (8.52).

The angles, β, β', θ, θ', δ_1', and δ_2' are connected by

$$\theta = \theta' - \varphi', \quad \beta' = \beta + \delta_1', \quad \theta = \delta_2' - \delta_1', \quad \delta_1' = \varphi', \quad \delta_2' = \theta' \tag{8.57a,b,c,d,e}$$

$$\tan\left(\delta_2' - \delta_1'\right) = \frac{1}{\tan\beta}\frac{Z}{\dfrac{\gamma+1}{2}m - Z} \tag{8.57f}$$

where $m = M_1^2$, and Z utilizes M_{1n}^2, not M_1^2. From these equations, we obtain

$$\beta_s = -\frac{d\varphi'}{ds} = -\frac{r_a \sin\beta'}{r^2} = -\left(r_a \sin\beta'\right)\frac{m}{X_1^{(\gamma+1)/(\gamma-1)}} \tag{8.58a}$$

$$\beta_s' = \beta_s + \frac{d\varphi'}{ds} = 0 \tag{8.58b}$$

The M_{1s} parameter is given by

$$M_{1s} = \frac{dM_1}{ds} = \frac{dM_1}{dr}\frac{dr}{d\varphi'}\frac{d\varphi'}{ds} = \frac{[m(m-w)]^{1/2}}{m-1}\frac{1}{X_1^{(3-\gamma)/[2(\gamma-1)]}}$$

or

$$\frac{M_{1s}}{M_1} = \frac{(m-w)^{1/2}}{m-1}\frac{1}{X_1^{(3-\gamma)/[2(\gamma-1)]}} \tag{8.59a}$$

From the isentropic relations for the pressure and density, we have

$$p_{1s} = -\frac{\gamma m}{X_1}\frac{M_{1s}}{M_1}, \quad \rho_{1s} = \frac{1}{\gamma}p_{1s} \tag{8.59b,c}$$

while Equation (8.14) yields

$$V_{1s} = \frac{1}{X_1}\frac{M_{1s}}{M_1} \tag{8.59d}$$

Note that M_1 and β, which appear in w, depend on φ', as will be evident in the subsequent example.

Conditions at state 2 are obtained from Appendix F with σ and β'_s equal to zero. For instance, for the pressure, we have

$$p_2 = f_3 = \frac{2}{\gamma+1}Y \tag{8.60a}$$

$$p_{2s} = g_{31}\,\chi_1 + g_{33}\,\chi_3 + g_{35}\,\chi_5 = \frac{4\gamma}{\gamma+1}wM_{1s} + \frac{2}{\gamma+1}Yp_{1s} + \frac{8\gamma}{(\gamma+1)^2}XA\beta_s \tag{8.60b}$$

$$p_{2n} = h_{31}\,\chi'_1 + h_{32}\,\chi'_2 + h_{33}\,\chi'_3 + h_{34}\,\chi'_4$$

$$= \frac{\gamma m\rho_2 v_2}{\Delta}\left[\frac{p_2}{m\rho_2} - (\gamma-1)u_2^2\right]u_{2s} - \frac{\gamma m\rho_2 u_2}{\Delta}\left[\frac{p_2}{m\rho_2} + (\gamma-1)v_2^2\right]v_{2s}$$

$$- \frac{(\gamma-1)u_2 v_2}{\Delta}p_{2s} + \frac{\gamma p_2 u_2 v_2}{\Delta\rho_2}\rho_{2s} \tag{8.60c}$$

Observe that β_s, in p_{2s} depends on the nondimensional length parameter $(r_a \sin\beta')$. With

$$\gamma = 1.4, \quad M_{1na} = 3, \quad \beta' = 60°, \quad \varphi' = 25° \tag{8.61}$$

prescribed, we evaluate the CST parameters, $(\partial p/\partial\tilde{s})_2$ and $(\partial\delta'_2/\partial\tilde{s})_2$, which are given by Equations (8.31) and (8.35), respectively. From Equations (8.49) and (8.50), we obtain

$$M_{1a} = 3.464, \quad r_a = 11.35, \quad r_a\sin\beta' = 9.826 \tag{8.62a}$$

We next obtain r, Equation (8.52), and M_1, Equation (8.44), as

$$r = 17.13, \quad M_1 = 3.912 \tag{8.62b}$$

where M_1 is closely estimated using standard gas dynamic tables for a nozzle's area ratio versus the Mach number. Other state 1 parameters are then readily evaluated:

$$M_{1n} = 2.244, \quad \delta_1' = 25°$$

$$m = 15.30, \quad w = 5.035, \quad X_1 = 4.061$$

$$X = 2.007, \quad Y = 6.849, \quad Z = 4.035, \quad A = 4.299, \quad B = 19.48 \quad (8.62c)$$

$$p_{1s} = -7.181 \times 10^{-2}, \quad \rho_{1s} = -5.129 \times 10^{-2}, \quad M_{1s} = 5.325 \times 10^{-2}$$

At the shock, we have

$$\beta = 35°, \quad \beta_s = -3.354 \times 10^{-2}, \quad \beta_s' = 0 \quad (8.62d)$$

while at state 2,

$$p_2 = 5.707, \quad \rho_2 = 3.010, \quad u_2 = 0.8192, \quad v_2 = 0.1905$$

$$p_{2s} = -0.8127, \quad \rho_{2s} = -0.2573, \quad u_{2s} = 2.198 \times 10^{-2}, \quad v_{2s} = -1.977 \times 10^{-3}$$

$$M_2 = 2.390, \quad \delta_2' = 46.91° \quad (8.62e)$$

$$\alpha_2 = 0.1446, \quad \alpha_3 = -5.021 \times 10^{-3}, \quad \Delta = 8.758 \times 10^{-2}$$

$$p_{2n} = -0.9179, \quad \rho_{2n} = -0.5562, \quad u_{2n} = -2.838 \times 10^{-2}, \quad v_{2n} = 8.323 \times 10^{-2}$$

$$\left(\frac{\partial p}{\partial \tilde{s}}\right)_2 = -0.9996, \quad \frac{\partial \delta_2'}{\partial \tilde{s}} = -1.557 \times 10^{-2} \quad (8.62f)$$

The last two parameters are the desired CST results. See Problem 22 for the vorticity.

With increasing distance along the shock, M_1 increases and p_1 and ρ_1 isentropically decrease, with the rate of decrease greater for the pressure. This is evident in the p_{1s} and ρ_{1s} values in the above list, whose ratio is γ. It is easy to demonstrate that M_{1n} decreases with s, and the shock strength therefore also decreases. Consequently, p_2 and ρ_2 have a much more rapid decrease with s than p_1 and ρ_1, as is evident from the p_{2s} and ρ_{2s} values.

The magnitudes of p_{2n} and ρ_{2n} are somewhat larger than their p_{2s} and ρ_{2s} counterparts. This stems, in part, from the angle between the state 2 streamline and the shock, $\beta' - \delta_2'$, being only 13.1°. The streamline is thus nearly parallel to the shock. The rate of change of p_2 and ρ_2 with s therefore results in larger magnitudes for p_{2n} and ρ_{2n}.

Upstream of the shock, the flow is expansive with $(\partial \delta_1' / \partial \tilde{s}) = 0$, because the streamlines are straight. The flow downstream of the shock is also expansive, as indicated by the negative value for p_{2n}. In turn, this results in a positive value for $(\partial p / \partial \tilde{n})_2$ and a slightly negative value for $(\partial \delta_2' / \partial \tilde{s})_2$. At this shock location, the streamline curves downward. Note that the streamline curves upward at state a.

The length normalization given by Equation (8.45) is arbitrary, because $\dot{m} / (\rho a)_o$ is unspecified. Hence, nondimensional derivative values, such as β_s, p_{1s}, and p_{2n}, are arbitrary to within an $\overline{R}$ multiplicative factor.

9 General Derivative Formulation

9.1 PRELIMINARY REMARKS

A different approach from that in Chapter 8 is used to extend the analysis to a three-dimensional shock whose upstream flow may be nonuniform. The distinction in the preceding chapter between angles based on $\overline{V}_1$ and the x-coordinate will no longer be necessary.

The quantities

$$F(x_i), v_{1,i}, p_1, \rho_1 \tag{9.1}$$

are presumed known functions of the x_i and are sufficiently differentiable, as needed. For Sections 9.3 through 9.8, these parameters are time independent; a steady flow is assumed. The last two sections, however, consider the unsteady case, and parameters, such as $v_{1,i}$, now explicitly depend on x_i and t. Knowledge of these items are the necessary and sufficient conditions for the analysis in this chapter. The above items parallel the approach discussed in Section 4.1 and should result in a computer-friendly set of algorithms. This approach provides a seamless transition from CFD or experimental data, in a laboratory frame, to the shock-based frame of the theory.

The next section is primarily devoted to establishing a general version of the $\hat{t}, \hat{n}, \hat{b}$ basis and related parameters. The third section establishes a three-dimensional shock test case that easily reduces to a two-dimensional or axisymmetric shock, where Chapters 4 through 6 and Appendix E apply. This is an elliptic paraboloid (EP) shock whose freestream is uniform. The EP model is systematically utilized to illustrate the steady theory, and to partially verify it. Sections 9.4 through 9.8 cover shock curvatures, vorticity, jump conditions and tangential derivatives, normal derivatives, and applications, respectively. Section 9.9 extends the analysis to the unsteady case. This analysis is applied in Section 9.10 to the unsteady, curved reflected shock observed in a shock tube experiment.

The material in Sections 9.2 through 9.7 is summarized in three successive appendices. The first, Appendix H, contains the general steady formulation, while the second lists equations when the freestream is uniform. The last of these appendices does the same but with the EP model. Appendix L summarizes the unsteady shock approach.

9.2   VECTOR RELATIONS

The shock's generic shape is written as

$$F = F(x_i) = 0 \tag{9.2}$$

As before, the normal to the surface is

$$\hat{n} = \frac{\nabla F}{|\nabla F|} \tag{9.3}$$

where

$$\nabla F = \sum F_{x_j}\, \hat{l}_j, \qquad |\nabla F| = \left(\sum F_{x_j}^2 \right)^{1/2}$$

Let $\vec{V}_1$ be the velocity just upstream of the shock:

$$\vec{V}_1 = \sum v_{1,j}\, \hat{l}_j \tag{9.4}$$

where the $v_{1,j}$ components are assumed to be known functions of the x_i that satisfy Equation (9.2). The arbitrary sign of F is chosen such that $\hat{n}$ points in the down-stream direction. This means the angle, β, in the flow plane is positive:

$$\sin\beta = \frac{\vec{V}_1}{V_1} \cdot \hat{n} > 0 \tag{9.5a}$$

where

$$V_1 = \left(\sum v_{1,j}^2 \right)^{1/2} \tag{9.6a}$$

The equation for $\sin\beta$ can be written as

$$\sin\beta = \frac{\sum v_{1,j} F_{x_j}}{V_1\,|\nabla F|} \tag{9.5b}$$

From Figure 9.1, $\vec{V}_1$ can also be written as

$$\frac{\vec{V}_1}{V_1} = \sin\beta\, \hat{n} + \cos\beta\, \hat{t} \tag{9.6b}$$

As before, a right-handed, orthonormal basis, $\hat{t}, \hat{n}, \hat{b}$, is introduced (see Figure 9.1). We proceed to write all vectors in terms of a Cartesian $\hat{l}_j$ basis, as is already the case for $\hat{n}$ and $\vec{V}_1$. From the figure, observe that

$$\hat{n} \times \frac{\vec{V}_1}{V_1} = -\hat{b}\sin(90 - \beta) = -\hat{b}\cos\beta \tag{9.7a}$$

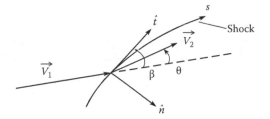

FIGURE 9.1 Shock-based basis and angles in the flow plane. $\hat{b}$ is into the page.

which becomes

$$\hat{b} = \frac{1}{\cos\beta} \frac{\vec{V}_1}{V_1} \times \hat{n} = -\chi \sum K_j \hat{l}_j \tag{9.7b}$$

where $\hat{b}$ and $\hat{l}_3$ have roughly opposite orientations, and

$$K_1 = v_{1,3} F_{x_2} - v_{1,2} F_{x_3}, \quad K_2 = v_{1,1} F_{x_3} - v_{1,3} F_{x_1}, \quad K_3 = v_{1,2} F_{x_1} - v_{1,1} F_{x_2} \tag{9.8}$$

$$\chi = \frac{1}{V_1 \, | \nabla F | \cos\beta} \tag{9.9}$$

The $\hat{t}$ vector is given by

$$\hat{t} = -\hat{b} \times \hat{n} = \frac{\chi}{|\Delta F|} \left(\sum K_j \hat{l}_j \right) \times \left(\sum F_{x_j} \hat{l}_j \right) = \frac{\chi}{|\nabla F|} \sum L_j \hat{l}_j \tag{9.10}$$

where

$$L_1 = F_{x_3} K_2 - F_{x_2} K_3, \quad L_2 = F_{x_1} K_3 - F_{x_3} K_1, \quad L_3 = F_{x_2} K_1 - F_{x_1} K_2 \tag{9.11a}$$

or

$$L_j = | \nabla F |^2 v_{1,j} - \left(\sum v_{1,k} F_{x_k} \right) F_{x_j} \tag{9.11b}$$

Since

$$\hat{n} \cdot \hat{b} = \hat{n} \cdot \hat{t} = \hat{t} \cdot \hat{b} = \frac{\vec{V}_1}{V_1} \cdot \hat{b} = 0, \quad \hat{b} \cdot \hat{b} = \hat{t} \cdot \hat{t} = 1 \tag{9.12}$$

we obtain

$$\sum F_{x_j} K_j = \sum F_{x_j} L_j = \sum K_j L_j = \sum v_{1,j} K_j = 0, \quad \sum K_j^2 = \frac{1}{\chi^2},$$

$$\sum L_j^2 = \frac{|\nabla F|^2}{\chi^2} \tag{9.13}$$

These sums can also be obtained by direct substitution of the K_j and L_j components. They are used to simplify some of the subsequent algebra.

This section concludes with relations for $\vec{V}_2$, where V_2 is given in Appendix E.1, by writing

$$\vec{V}_2 = V_2 \hat{\vec{s}} \tag{9.14}$$

The state 2 vector, $\hat{\vec{s}}$, is evaluated using

$$\hat{\vec{s}} = \frac{\vec{V}_2}{V_2} = a\,\hat{n} + c\,\frac{\vec{V}_1}{V_1}$$

With the aid of

$$\frac{\vec{V}_2}{V_2} \cdot \frac{\vec{V}_1}{V_1} = \cos\theta, \quad \hat{n} \cdot \frac{\vec{V}_1}{V_1} = \sin\beta$$

the a and c coefficients are established. These equations provide

$$a\,\sin\beta + c = \cos\theta, \qquad a\,\sin(\beta - \theta) + c\,\cos\theta = 1$$

which results in

$$\hat{\vec{s}} = -\frac{\sin\theta}{\cos\beta}\,\hat{n} + \frac{\cos(\beta - \theta)}{\cos\beta}\,\frac{\vec{V}_1}{V_1} \tag{9.15a}$$

Alternative forms are

$$\hat{\vec{s}} = \sin(\beta - \theta)\hat{n} + \cos(\beta - \theta)\hat{t} = \frac{1}{B^{1/2}}\left(\hat{n} + A\hat{t}\right) \tag{9.15b}$$

where β and θ are measured relative to $\vec{V}_1$, as indicated in Figure 9.1.

Appendices H.1, I.1, and J.1 summarize the foregoing, where Appendix J.1.1 utilizes the EP model of the next section.

9.3 ELLIPTIC PARABOLOID SHOCK

Before introducing the EP shock equation, the simplification associated with a uniform freestream

$$\vec{V}_1 = V_1 \hat{1}_1 = v_{1,1} \hat{1}_1, \quad v_{1,2} = v_{1,3} = 0, \quad V_1 = constant \tag{9.16}$$

is discussed. The sine and cosine of β become

$$\sin\beta = \frac{F_{x_1}}{|\nabla F|}, \quad \cos\beta = \frac{\psi^{1/2}}{|\nabla F|}$$

where

$$\psi = F_{x_2}^2 + F_{x_3}^2$$

Other parameters simplify to

$$K_1 = 0, \quad K_2 = V_1 F_{x_3}, \quad K_3 = -V_1 F_{x_2}$$

$$L_1 = V_1 \psi, \quad L_2 = -V_1 F_{x_1} F_{x_2}, \quad L_3 = -V_1 F_{x_1} F_{x_2}$$

and, from Equation (9.13),

$$\chi = \frac{1}{\left(\sum K_j^2\right)^{1/2}} = \frac{1}{V_1 \psi^{1/2}}$$

The $\hat{n}$ basis vector is unchanged, but $\hat{b}$ and $\hat{t}$ become

$$\hat{b} = \frac{1}{\psi^{1/2}}\left(-F_{x_3} \hat{1}_2 + F_{x_2} \hat{1}_3\right)$$

$$\hat{t} = \frac{1}{|\nabla F| \psi^{1/2}}\left[\psi \hat{1}_1 - F_{x_1}\left(F_{x_2} \hat{1}_2 + F_{x_3} \hat{1}_3\right)\right]$$

Although the shock is three-dimensional, because $(\vec{V}_1 / V_1) = \hat{1}_1$ and $\hat{b}$ is normal to the flow plane, $\hat{b}$ does not have a component along $\hat{1}_1$.

This is also a convenient time to discuss some differences between a three-dimensional shock and a two-dimensional or axisymmetric shock. These differences focus on intrinsic coordinates, previously discussed in Section 5.2.

For a convex shock, $\hat{\tilde{n}}_2$ points into the upstream flow. The binormal unit basis, $\hat{\tilde{b}}$, is normal to the streamline's osculating plane and is given by

$$\hat{\tilde{b}} = \hat{\tilde{n}} \times \hat{\tilde{s}} \tag{9.17}$$

As with $\hat{b}$, $\hat{\tilde{b}}$ points into a page that contains $\hat{\tilde{n}}$ and $\hat{\tilde{s}}$.

An intrinsic coordinate system is definable at any point in the flow, including at state 2. (A uniform flow is an exception.) As previously mentioned, when a shock is two-dimensional or axisymmetric, the osculating and flow planes coincide, and $\hat{\tilde{b}} = \hat{b}$. When a shock is three-dimensional, however, the two planes do not coincide, except along $\hat{\tilde{s}}$. In the three-dimensional case, the osculating plane is rotated from the flow plane along a line containing $\hat{\tilde{s}}$, and the vectors $\hat{b}$ and $\hat{\tilde{b}}$ are no longer collinear. Except when the shock is a normal shock, any vector normal to $\hat{\tilde{s}}$, or $\vec{V}_2$, as is $\hat{\tilde{b}}$ and $\hat{\tilde{n}}$, cannot be in the shock's tangential plane.

The scalar momentum equations are simple and elegant when written with intrinsic coordinates. Of special interest is the steady momentum equation that is normal to the osculating plane (Serrin 1959):

$$\frac{\partial p}{\partial \tilde{b}} = 0 \tag{9.18}$$

Since coordinates b and $\tilde{b}$ do not coincide, this implies that $(\partial p/\partial b) \neq 0$ at state 2. This is demonstrated shortly using an EP shock. In addition, Appendix H.4 provides relations for the nonzero b derivatives, including for the pressure. The distinction between the flow and osculating planes and Equation (9.18) are of considerable importance in Section 9.7, where the n derivatives are evaluated.

It is convenient for a three-dimensional shock to summarize the above discussion concerning the b and $\tilde{b}$ coordinates, and to discuss a third coordinate, b_{sh}, which is introduced in Section 9.8. All three coordinates are distinct, originate at the same state 2 location, and are only locally defined. The b coordinate is in the shock's tangent plane, is normal to the flow plane, and, for example, $(\partial p/\partial b)_2$ is not zero. The $\tilde{b}$ coordinate is normal to the osculating plane and has $\partial p/\partial \tilde{b}$ equal to zero; it is not in the tangent plane but is in the $\tilde{s}, b_{sh}$ plane. The b_{sh} coordinate is defined as being in the tangent plane with $\partial p/\partial b_{sh} = 0$. It is not in the osculating plane or normal to either the flow or osculating planes.

The chosen EP shock shape is

$$F = x_1 - \frac{x_2^2}{2r_2} - \frac{\sigma x_3^2}{2r_3} = 0 \tag{9.19}$$

where r_2 is the radius of curvature at the nose in the $x_3 = 0$ plane, and r_3 is the radius of curvature at the nose in the $x_2 = 0$ plane.

The shock is convex relative to the upstream flow and is three-dimensional when $r_2 \neq r_3$. It is two-dimensional when $\sigma = 0$ and axisymmetric when

$$\sigma = 1, \qquad r = r_2 = r_3, \qquad y^2 = x_2^2 + x_3^2 \tag{9.20a}$$

For both the two-dimensional and axisymmetric cases, it has the form of Equation (3.2) if f is written as

$$f = (2\, r_2\, x_1)^{1/2} = (2\, r\, x_1)^{1/2} \tag{9.20b}$$

The orientation of an isobaric (constant p_2) curve through a fixed x_i point on an EP shock is evaluated. From Appendix E.1, this corresponds to a constant value for β. Moreover, this curve also has constant values for ρ_2, u_2, v_2, ... A constant β value results in (see Equation J.4)

$$\psi = \left(\frac{x_2}{r_2}\right)^2 + \left(\frac{\sigma x_3}{r_3}\right)^2 = constant \tag{J.3}$$

A tangent vector to the isobaric curve on the shock's surface is given by

$$\vec{a} = \nabla F \times \nabla \psi$$

where

$$\nabla \psi = \frac{2\, x_2}{r_2^2}\, \hat{i}_2 + \frac{2\, \sigma x_3}{r_3^2}\, \hat{i}_3$$

Hence, $\vec{a}$ equals

$$\vec{a} = 2\left[\frac{\sigma x_2 x_3}{r_2 r_3}\left(\frac{1}{r_2} - \frac{1}{r_3}\right)\hat{i}_1 - \frac{\sigma x_3}{r_3^2}\,\hat{i}_2 + \frac{x_2}{r_2^2}\,\hat{i}_3\right]$$

whereas $\hat{b}$ is given by

$$\hat{b} = \frac{1}{\psi^{1/2}}\left(\frac{\sigma x_3}{r_3}\,\hat{i}_2 - \frac{x_2}{r_2}\,\hat{i}_3\right) \tag{J.10}$$

The two vectors are collinear only when one of the Equation (G.22) (Appendix G), array is satisfied. Hence, when a shock is three-dimensional, the isobaric curve is not tangent to the b-coordinate.

9.4 SHOCK CURVATURES

For consistency with curved shock theory (Molder 2012), the earlier curvatures, $-d\beta/ds$ and $\sigma\cos\beta/y$, are replaced with S_a and S_b, respectively. The S_a curvature is in the flow plane, while S_b is in the plane that is normal to both the shock and flow plane. As before, both curvatures are negative (positive) when the shock is concave (convex) relative to the upstream flow. If the point of interest is in the vicinity of a saddle point, the curvatures are of opposite signs. At a saddle point, both curvatures have extremum values of the opposite sign.

Each curvature is for a curve at a prescribed point, x_i^*, on the shock, and the subsequent analysis requires a Cartesian basis, $\hat{l}_i$ defined below, and a corresponding coordinate system, x_i'. This coordinate system is required because the curve, whose curvature is S_m, $m = a, b$, is written in terms of the x_1' and x_2' coordinates.

For S_a, the right-handed system

$$\hat{l}_1 = \hat{n}, \quad \hat{l}_2 = \hat{t}, \quad \hat{l}_3 = -\hat{b}$$

is chosen such that x_1', x_2' are in the flow plane. The subsequent analysis makes use of the differential operator approach in Appendix G. From Equation (G.17a,c), the linear, differential operators are written as

$$\frac{\partial}{\partial x_1'} = \frac{1}{|\nabla F|^*} \sum F_{x_j}^* \frac{\partial}{\partial x_j} \tag{9.21a}$$

$$\frac{\partial}{\partial x_2'} = \frac{\chi^*}{|\nabla F|^*} \sum L_j^* \frac{\partial}{\partial x_j} \tag{9.22a}$$

where the asterisk indicates evaluation at a fixed shock point, x_i^*, where the coefficients of the $\partial()/\partial x_j$, such as $F_{x_j}^*/|F|^*$, are constants. The "a" used with the equation numbering indicates these operators are associated with S_a.

For S_b, the right-handed system ($\hat{b}$ points into the page)

$$\hat{l}_1 = \hat{n}, \quad \hat{l}_2 = \hat{b}, \quad \hat{l}_3 = \hat{t}$$

is chosen such that x_1', x_2' are in the $\hat{n}, \hat{b}$ or S_b plane. From Equation (G.17b,c), the operators are

$$\frac{\partial}{\partial x_1'} = \frac{1}{|\nabla F|^*} \sum F_{x_j}^* \frac{\partial}{\partial x_j} \tag{9.23b}$$

$$\frac{\partial}{\partial x_2'} = -\chi^* \sum K_j^* \frac{\partial}{\partial x_j} \tag{9.24b}$$

where, again, the coefficients of $\partial()/\partial x_j$ are constants evaluated at x_i^*.

These operators generate the derivatives

$$F_{x_1'}, \quad F_{x_2'}, \quad F_{x_1'x_1'}, \quad F_{x_1'x_2'} = F_{x_2'x_1'}, \quad F_{x_2'x_2'} \tag{9.25}$$

required for S_m, $m = a, b$. The F, of course, is given by Equation (9.2), but F_{xy} is given by Equation (9.22a), for S_a, while Equation (9.24b) is used with S_b. The relevant implicit curvature equation (Courant 1949) is

$$S_m = -\frac{F_{x_2}^2 F_{x_1 x_1} - 2F_{x_1} F_{x_2} F_{x_1 x_2} + F_{x_1}^2 F_{x_2 x_2}}{\left(F_{x_1}^2 + F_{x_2}^2\right)^{3/2}}, \quad m = a, b \tag{9.26}$$

After the F derivatives are evaluated, the x_i^* are no longer required and are replaced with x_i. Thus, S_m only depends on the now variable shock point, x_i. This curvature relation, as written, is in accord with the earlier S_m sign convention.

The EP shock yields (see Appendix J.2)

$$S_a = \frac{\dfrac{x_2^2}{r_2^3} + \dfrac{\sigma x_3^2}{r_3^3}}{\left[\left(\dfrac{x_2}{r_2}\right)^2 + \left(\dfrac{\sigma x_3}{r_3}\right)^2\right]\left[1 + \left(\dfrac{x_2}{r_2}\right)^2 + \left(\dfrac{\sigma x_3}{r_3}\right)^2\right]^{3/2}} \tag{9.27a}$$

$$S_b = \frac{\sigma\left(\dfrac{x_2^2}{r_2} + \dfrac{x_3^2}{r_3}\right)}{r_2 r_3 \left[\left(\dfrac{x_2}{r_2}\right)^2 + \left(\dfrac{x_3}{r_3}\right)^2\right]\left[1 + \left(\dfrac{x_2}{r_2}\right)^2 + \left(\dfrac{x_3}{r_3}\right)^2\right]^{1/2}} \tag{9.27b}$$

When the flow is two-dimensional, this becomes

$$S_a = \frac{1}{r_2\left[1 + \left(\dfrac{x_2}{r_2}\right)^2\right]^{3/2}}, \quad S_b = 0 \tag{9.28a,b}$$

This is readily confirmed for S_a using the explicit curvature formula

$$\kappa = -\frac{\dfrac{d^2 x_2}{dx_1^2}}{\left[1 + \left(\dfrac{dx_2}{dx_1}\right)^2\right]^{3/2}} \tag{9.29}$$

with $x_2 = (2r_2 x_1)^{1/2}$. (The curvature sign convention adopted at the start of Section 5.1 requires the minus sign.) Note that $S_a = r_2^{-1}$ when $x_2 = 0$, as expected.

When the shock is axisymmetric, Equation (9.20a) applies, with the result

$$S_a = \frac{r^2}{\left(r^2 + y^2\right)^{3/2}}, \quad S_b = \frac{1}{\left(r^2 + y^2\right)^{1/2}} \tag{9.30a,b}$$

Again, both curvatures become r^{-1} when $y = 0$. The S_a result agrees with Equation (9.28a) when $r_2 = r$, and S_b agrees with

$$S_b = \frac{\cos\beta}{y} \tag{9.31}$$

with the $\cos\beta$ relation in Equation (J.4).

9.5 VORTICITY

In Chapter 6, the homenergetic condition is automatic, because the upstream flow is uniform and the shock is steady. In Section 8.6, this requirement is specifically not assumed, because the flow is two-dimensional or axisymmetric. We now discuss its relevance when the shock's shape is three-dimensional.

The vorticity, $\vec{\omega}_2$, can also be obtained with the curvilinear curl operating on the flow-plane velocity using shock-based coordinates. The uniform upstream flow result (see Equation 6.8c) contains a normal derivative term that must be deleted if the well-known shock-generated vorticity equation is to be obtained. In this circumstance, $\vec{\omega}_2$ is tangential to the shock.

The requirement of a homenergetic flow in the above tangency discussion can be demonstrated with a two-dimensional flow with sweep (Emanuel 2001, Chapter 10). The flow initially consists of two uniform, supersonic flow regions separated by a planar, oblique shock, where the shock has a sweep angle (i.e., the attached shock is generated by a straight ramp with sweep). This flow, of course, is irrotational, homenergetic, and the streamlines are straight. By means of the substitution principle (Emanuel 2001, Chapter 8), the two uniform flows are transformed into nonuniform, parallel flows. The substitution principle generates a new rotational, nonhomenergetic flow that still satisfies the steady, three-dimensional Euler equations and leaves the geometry unchanged. Thus, the straight streamlines and the planar shock, with sweep, are invariant. The vorticity vector is perpendicular to the straight streamlines at states 1 and 2 and is *not* tangent to the shock. The difference between the two flows is that the first is homenergetic, the second is not. The nontangency result is also apparent from Crocco's equation, Equation (6.2a), which contains the gradient of the stagnation enthalpy. In this section, we thus assume a homenergetic flow in order that $\vec{\omega}_1$ and $\vec{\omega}_2$ are tangent to the shock.

Derivatives of parameters, such as M_1, are in terms of the x_i. For instance, for M_1, we have

$$M_1^2 = \frac{\rho_1 V_1^2}{\gamma p_1} \tag{9.32}$$

and, consequently,

$$\frac{M_{1x_i}}{M_1} = \frac{1}{V_1^2} \sum v_{1,j} v_{1,jx_i} - \frac{1}{2} \frac{p_{1x_i}}{p_1} + \frac{1}{2} \frac{\rho_{1x_i}}{\rho_1} \tag{9.33}$$

Another derivative parameter, encountered shortly, is β_{x_i}. We start with Equation (9.5b) and

$$\cos\beta = \frac{1}{V_1 \, |\nabla F| \, \chi} \tag{9.34}$$

to obtain

$$\beta_{x_i} = \frac{\chi}{V_1^2 \, |\nabla F|^2} \left\{ |\nabla F|^2 \left[V_1^2 \sum v_{1,jx_i} F_{x_j} - \left(\sum v_{1,j} F_{x_j} \right) \sum v_{1,j} v_{1,jx_i} \right] \right.$$

$$\left. + V_1^2 \left[|\nabla F|^2 \sum v_{1,j} F_{x_i x_j} - \left(\sum v_{1,j} F_{x_j} \right) \sum F_{x_j} F_{x_i x_j} \right] \right\} \tag{9.35}$$

where j is summed over in each summation. Thus, β_{x_i} is fully determined by F and $v_{1,j}$ and derivatives thereof.

Some of the subsequent discussion follows that in Section 8.6. It is convenient to write the velocity, vorticity, and gradient operator as

$$\vec{V} = u\hat{t} + v\hat{n} \tag{9.36a}$$

$$\vec{\omega} = \omega_b \hat{b} + \omega_t \hat{t} \tag{9.36b}$$

$$\nabla = \hat{t}\frac{\partial}{\partial s} + \hat{n}\frac{\partial}{\partial n} + \hat{b}\frac{\partial}{\partial b} \tag{9.36c}$$

Crocco's equation results in

$$T\frac{\partial S}{\partial b} = v\omega_t, \quad T\frac{\partial S}{\partial s} = -v\omega_b, \quad T\frac{\partial S}{\partial n} = u\omega_b \tag{9.37a,b,c}$$

Equations (9.37b,c) yield

$$u\frac{\partial S}{\partial s} + v\frac{\partial S}{\partial n} = 0$$

which is the equation for an isentropic flow. Equation (9.37a,b) is written as

$$\frac{\partial S}{\partial b} = \frac{v}{T}\omega_t, \quad \frac{\partial S}{\partial s} = -\frac{v}{T}\omega_b \tag{9.38a,b}$$

The entropy change across a shock is

$$S_2 - S_1 = \frac{R}{\gamma - 1} \ln\left[\left(\frac{\gamma + 1}{2}\right)^{(\gamma + 1)} \frac{w^\gamma}{X^\gamma Y}\right] \tag{9.39}$$

The entropy jump and w equations are differentiated with the result

$$\frac{\partial S_2}{\partial x_i} = \frac{\partial S_1}{\partial x_i} + \frac{\gamma R}{2} \frac{Z^2}{wXY} w_{x_i} \tag{9.40}$$

$$w_{x_i} = 2w\left[\frac{M_{1x_i}}{M_1} + (\cot\beta)\beta_{x_i}\right] \tag{9.41}$$

where M_{1x_i} and β_{x_i} are given by Equations (9.33) and (9.35), respectively. Elimination of w_{x_i} yields

$$\frac{\partial S_2}{\partial x_i} = \frac{\partial S_1}{\partial x_i} + \gamma R \frac{Z^2}{XY}\left[\frac{M_{1x_i}}{M_1} + (\cot\beta)\beta_{x_i}\right] \tag{9.42}$$

The $\partial S/\partial b$ derivative is related to the $\partial S/\partial x_i$ derivatives by Equation (G.17b), with the result

$$\frac{\partial S_1}{\partial b} = -\chi \sum K_i \frac{\partial S_1}{\partial x_i} \tag{9.43}$$

$$\frac{\partial S_2}{\partial b} = -\chi \sum K_i \frac{\partial S_2}{\partial x_i}$$

$$= -\chi \sum K_i \frac{\partial S_1}{\partial x_i} - \gamma R \frac{Z^2\chi}{XY} \sum K_i\left[\frac{M_{1x_i}}{M_1} + (\cot\beta)\beta_{x_i}\right]$$

$$= \frac{\partial S_1}{\partial b} - \gamma R \frac{Z^2\chi}{XY} \sum K_i\left[\frac{M_{1x_i}}{M_1} + (\cot\beta)\beta_{x_i}\right] \tag{9.44}$$

This is combined with Equation (9.38a) to obtain

$$\frac{v_2}{T_2}\omega_{t2} = \frac{v_1}{T_1}\omega_{t1} - \gamma R \frac{Z^2\chi}{XY} \sum K_i\left[\frac{M_{1x_i}}{M_1} + (\cot\beta)\beta_{x_i}\right] \tag{9.45}$$

With Equations (G.17a), (9.38b), and (9.42), the other vorticity component is

$$\frac{v_2}{T_2}\,\omega_{b2} = \frac{v_1}{T_1}\,\omega_{b1} - \gamma R \frac{Z^2\chi}{XY\,|\nabla F|}\sum L_i\left[\frac{M_{1x_i}}{M_1} + (\cot\beta)\beta_{x_i}\right] \qquad (9.46)$$

With the following relations

$$v_1 = V_1\sin\beta\,, \qquad v_2 = V_2\sin(\beta-\theta)\,, \qquad M_1^2 = \frac{V_1^2}{\gamma R T_1}$$

$$\frac{T_2}{T_1} = \left(\frac{2}{\gamma+1}\right)^2\frac{XY}{w}\,, \qquad \frac{p_2}{p_1} = \frac{2}{\gamma+1}Y \qquad (9.47)$$

$$\frac{V_2}{V_1} = \left[\cos^2\beta + \left(\frac{2}{\gamma+1}\right)^2\frac{X^2\sin^2\beta}{w^2}\right]^{1/2} = \frac{2}{\gamma+1}\frac{X}{w}\frac{\sin\beta}{\sin(\beta-\theta)}$$

the components are given by

$$\omega_{b2} = \frac{p_2}{p_1}\,\omega_{b1} - Q_b \qquad (9.48)$$

$$\omega_{t2} = \frac{p_2}{p_1}\,\omega_{t1} - Q_t \qquad (9.49)$$

where

$$Q_b = \frac{2}{\gamma+1}\frac{Z^2}{wX}\frac{\tan\beta}{|\nabla F|^2}\sum L_i\left[\frac{M_{1x_i}}{M_1} + (\cot\beta)\beta_{x_i}\right] \qquad (9.50a)$$

$$Q_t = \frac{2}{\gamma+1}\frac{Z^2}{wX}\frac{\tan\beta}{|\nabla F|}\sum K_i\left[\frac{M_{1x_i}}{M_1} + (\cot\beta)\beta_{x_i}\right] \qquad (9.50b)$$

The vorticity then is

$$\vec{\omega}_2 = \frac{p_2}{p_1}\,\vec{\omega}_1 - Q_b\hat{b} - Q_t\hat{t} \qquad (9.51a)$$

and its magnitude is

$$\omega_2 = \pm\left[\left(\frac{p_2}{p_1}\omega_{b1} - Q_b\right)^2 + \left(\frac{p_2}{p_1}\omega_{t1} - Q_t\right)^2\right]^{1/2} \qquad (9.51b)$$

where the plus (minus) sign is used for a convex (concave) shock. Other than being tangent to the shock, the orientation of $\vec{\omega}_2$ is not simply related to $\vec{\omega}_1$ or to either curvature plane.

For a steady, homenergetic flow, $\vec{\omega}_2$ is determined by the upstream vorticity, amplified by p_2/p_1, the gradient of M_1 on the shock's surface, and the β_{x_i} factor. In particular, the M_1 and β gradient terms account for any irrotational upstream nonuniformity not associated with the upstream vorticity.

Appropriate summaries are provided in Appendices H, I, and J. Equations (J.15), (J.16), and (J.17) provide the EP shock results for Q_b, Q_t, and $\vec{\omega}_2$. The sign of Q_t depends on the relative size of the r_i. When $r_2 > r_3$, Q_t and Q_b are both negative for a convex shock. When the shock is axisymmetric, $Q_t = 0$ and Q_b has its conventional form:

$$Q_b = \frac{2}{\gamma+1} \frac{Z^2}{wX} V_1 \left(\cos\beta\right) \beta_s \qquad (9.52)$$

where, with the aid of Equation (G.17a), we obtain

$$\beta_s = \frac{\partial\beta}{\partial s} = \sum \beta_{x_i} \frac{\partial x_i}{\partial s} = \frac{\chi}{|\nabla F|} \sum L_i \beta_{x_i} \qquad (9.53)$$

9.6 JUMP CONDITIONS AND TANGENTIAL DERIVATIVES

Appendices E.1 or F.1 can be used for the jump conditions. Our next goal is to formulate equations for the s and b derivatives of u, v, p, and ρ, at state 2, in terms of the known (Equation 9.1) parameters and their x_i derivatives. Assisting in this effort are Equations (9.5b) and (9.35) for β and β_{x_i} and Equations (9.32) and (9.33) for M_1 and M_{1x_i}. The connection between the two types of derivatives is provided by

$$\frac{\partial}{\partial s} = \frac{\chi}{|\nabla F|} \sum L_i \frac{\partial}{\partial x_i} \qquad (G.17a)$$

$$\frac{\partial}{\partial b} = -\chi \sum K_i \frac{\partial}{\partial x_i} \qquad (G.17b)$$

In using Appendix E.1, w (Equation 4.2) is often encountered, whose state 1 surface derivatives are

$$\frac{1}{w} \frac{\partial w}{\partial s} = 2\frac{\chi}{|\nabla F|} \left(\sum L_i \frac{M_{1x_i}}{M_1} + \cot\beta \sum L_i \beta_{x_i} \right) \qquad (9.54a)$$

$$\frac{1}{w} \frac{\partial w}{\partial b} = -2\chi \left(\sum K_i \frac{M_{1x_i}}{M_1} + \cot\beta \sum K_i \beta_{x_i} \right) \qquad (9.54b)$$

This operator approach is illustrated by obtaining the pressure derivatives, where

$$p_2 = \frac{2}{\gamma+1} p_1 Y \tag{9.55}$$

We thus have

$$\frac{1}{p_1}\left(\frac{\partial p}{\partial s}\right)_2 = \frac{2}{\gamma+1}\left(\frac{Y\chi}{|\nabla F|}\sum L_i \frac{p_{1x_i}}{p_1} + \gamma \frac{\partial w}{\partial s}\right) \tag{G.20}$$

$$= \frac{2}{\gamma+1}\frac{\chi}{|\nabla F|}\left(Y\sum L_i \frac{p_{x_i}}{p_1} + 2\gamma w \sum L_i \frac{M_{1x_i}}{M_1} + 2\gamma w \cot\beta \sum L_i \beta_{x_i}\right) \tag{9.56}$$

$$\frac{1}{p_1}\left(\frac{\partial p}{\partial b}\right)_2 = -\frac{2}{\gamma+1}\chi\left(Y\sum K_i \frac{p_{x_i}}{p_1} + 2\gamma w \sum K_i \frac{M_{1x_i}}{M_1} + 2\gamma w \cot\beta \sum K_i \beta_{x_i}\right)$$

$$\tag{9.57}$$

which are Equations (H.17) and (H.18).

Appendix H.3 summarizes the s and b tangential derivatives, while simplified versions are provided in Appendix I.3 and Appendix J.3. With the u, v, p, and ρ derivatives available, other derivatives, such as for T_2 or M_2, can be obtained.

The s derivatives are checked against Appendix E.2 when the freestream is uniform and the shock is two-dimensional or axisymmetric. The comparison is expedited with the use of Equation (9.53). For example, the Appendix E.2 equation for $(\partial p/\partial s)_2$ becomes (with $\beta' = \beta_s$)

$$\frac{1}{p_1}\left(\frac{\partial p}{\partial s}\right)_2 = \frac{4}{\gamma+1}\beta'm \sin\beta \cos\beta = \frac{4\gamma}{\gamma+1}w \cot\beta \frac{\chi}{|\nabla F|}\sum L_i \beta_{x_i} \tag{9.58}$$

which agrees with Equation (H.17) when p_{x_i} and M_{x_i} are zero.

One can show that $p_{2s}[= (\partial p/\partial s)_2/p_1]$ in Appendix F, where the upstream flow is nonuniform, is in accord with Equation (9.56). To obtain agreement, use Equations (9.53), (G.18), and (9.54a). We also note that from Equation (J.27) that $(\partial p/\partial b)_2$ is zero if $\sigma = 0$ or $r_2 = r_3$.

9.7 NORMAL DERIVATIVES

When the normal derivative analysis was started, it was expected that the two-dimensional/axisymmetric approach in Chapters 3 and 4 would suffice. After much effort, it was realized that a global, shock-based coordinate system, ξ_i, as determined in Chapter 3, does not exist for a three-dimensional shock. This demonstration is relegated to Appendix K. In the following, the steady Euler equations are directly obtained in terms of s, n, and b coordinates. The corresponding h_i scale factors appear only in logarithmic derivative form. A variety of techniques are used to evaluate these derivatives, with a number of them established with the aid of the operator approach in Appendix G. Algebraic equations for the scale factors are not obtained and integration of these derivatives is not necessary, and, in fact, is not possible. The h_i derivatives are evaluated in terms of the x_i-coordinates; a relation between these coordinates and the s,n,b coordinates, however, does not exist. As will become apparent, the method is quite different from that in Chapter 3 and Section 4.4.

The velocity and gradient operator are written as

$$\bar{V} = u\hat{t} + v\hat{n} + \bar{w}\hat{b} \tag{9.59}$$

$$\nabla = \hat{t}\,\frac{\partial}{\partial s} + \hat{n}\,\frac{\partial}{\partial n} + \hat{b}\,\frac{\partial}{\partial b} \tag{9.60}$$

where, at state 2,

$$\bar{w} = 0\,, \quad \frac{\partial \bar{w}}{\partial s} = \frac{\partial \bar{w}}{\partial b} = 0\,, \quad \frac{\partial \bar{w}}{\partial n} \neq 0 \tag{9.61}$$

The $\hat{b}$ velocity component, $\bar{w}$, has the form $\bar{w} = \bar{w}(n)$, $\bar{w}(0) = 0$. Of course, for a two-dimensional or axisymmetric shock, $\partial \bar{w}/\partial n$ is also zero. In the general case, a streamline, at state 2, generally has a nonzero torsion value. It is essential that $\partial \bar{w}/\partial n$ not be zero if torsion is to occur. Although $\bar{V}_2$ is in the flow plane, and the streamline is tangent to the flow plane at state 2, because $\partial p/\partial n$ is not zero, the streamline curves away from the flow plane (i.e., it has nonzero torsion).

Kaneshige and Hornung (1999), in a correction to an earlier paper, and Hornung (2010) provide a formulation for the normal derivatives of a three-dimensional shock. A flow plane is utilized and $\bar{w}$ and its flow plane derivatives, including $\partial \bar{w}/\partial n$, are taken as zero. Only the derivative of $\bar{w}$ normal to the flow plane is nonzero. The derivatives of p, ρ, and u that are normal to the flow plane are also taken as zero. This formulation disagrees with Equations (9.18) and (9.61). The impact of the osculating plane, as discussed in the EP shock section, has been overlooked. In particular, the derivative of p that is normal to the osculating plane is zero. It is not zero when normal to the flow plane.

At state 2, the unknowns to be evaluated are

$$\frac{\partial u}{\partial n}\,, \quad \frac{\partial v}{\partial n}\,, \quad \frac{\partial p}{\partial n}\,, \quad \frac{\partial \rho}{\partial n}\,, \quad \frac{\partial \bar{w}}{\partial n} \tag{9.62}$$

For these parameters, the five governing scalar equations are continuity, momentum (three scalar equations), and an isentropic flow equation:

$$\rho \nabla \cdot \bar{V} + \bar{V} \cdot \nabla \rho = 0 \tag{9.63a}$$

$$\frac{D\bar{V}}{Dt} + \frac{1}{\rho}\nabla p = 0 \tag{9.63b,c,d}$$

$$\bar{V} \cdot \nabla\left(\frac{p}{\rho^{\gamma}}\right) = 0 \tag{9.63e}$$

The isentropic relation is equivalent to but somewhat simpler than a stagnation enthalpy relation, as previously used in Section 4.4. In the following discussion, the acceleration and then the divergence of the velocity are first evaluated with shock-based coordinates.

For a steady flow, Problem 7 provides

$$
\frac{D\vec{V}}{Dt} = \vec{V} \cdot \left(\nabla \vec{V}\right) = u\frac{\partial u}{\partial s}\,\hat{t} + u^2\frac{\partial \hat{t}}{\partial s} + u\frac{\partial v}{\partial s}\,\hat{n} + uv\frac{\partial \hat{n}}{\partial s} + u\frac{\partial \overline{w}}{\partial s}\,\hat{b} + u\overline{w}\frac{\partial \hat{b}}{\partial s}
$$

$$
+ v\frac{\partial u}{\partial n}\,\hat{t} + uv\frac{\partial \hat{t}}{\partial n} + v\frac{\partial v}{\partial n}\,\hat{n} + v^2\frac{\partial \hat{n}}{\partial n} + v\frac{\partial \overline{w}}{\partial n}\,\hat{b} + v\overline{w}\frac{\partial \hat{b}}{\partial n} + \overline{w}\frac{\partial u}{\partial b}\,\hat{t}
$$

$$
+ u\overline{w}\frac{\partial \hat{t}}{\partial b} + \overline{w}\frac{\partial v}{\partial b}\,\hat{n} + v\overline{w}\frac{\partial \hat{n}}{\partial b} + \overline{w}\frac{\partial \overline{w}}{\partial b}\,\hat{b} + \overline{w}^2\frac{\partial \hat{b}}{\partial b}
\tag{9.64a}
$$

With Equation (9.61), this simplifies to

$$
\vec{V} \cdot \left(\nabla \vec{V}\right) = u\frac{\partial u}{\partial s}\,\hat{t} + u^2\frac{\partial \hat{t}}{\partial s} + u\frac{\partial v}{\partial s}\,\hat{n} + uv\frac{\partial \hat{n}}{\partial s} + v\frac{\partial u}{\partial n}\,\hat{t} + uv\frac{\partial \hat{t}}{\partial n}
$$

$$
+ v\frac{\partial v}{\partial n}\,\hat{n} + v^2\frac{\partial \hat{n}}{\partial n} + v\frac{\partial \overline{w}}{\partial n}\,\hat{b}
\tag{9.64b}
$$

The use of Equation (9.61) results in the absence of b derivatives in Equation (9.64b). In the subsequent analysis, the s tangential derivatives for u, v, p, and ρ are required but not the corresponding b tangential derivatives, except for the pressure.

To eliminate the basis derivative terms, Equations (3.10), (3.31), and (3.37) are utilized, with the result

$$
\frac{\partial \hat{t}}{\partial s} = -\frac{1}{h_1}\frac{\partial h_1}{\partial n}\,\hat{n} - \frac{1}{h_1}\frac{\partial h_1}{\partial b}\,\hat{b}, \quad \frac{\partial \hat{t}}{\partial n} = \frac{1}{h_2}\frac{\partial h_2}{\partial s}\,\hat{n}, \quad \frac{\partial \hat{t}}{\partial b} = \frac{1}{h_3}\frac{\partial h_3}{\partial s}\,\hat{b}
$$

$$
\frac{\partial \hat{n}}{\partial s} = \frac{1}{h_1}\frac{\partial h_1}{\partial n}\,\hat{t}, \quad \frac{\partial \hat{n}}{\partial n} = -\frac{1}{h_2}\frac{\partial h_2}{\partial s}\,\hat{t} - \frac{1}{h_2}\frac{\partial h_2}{\partial b}\,\hat{b}, \quad \frac{\partial \hat{n}}{\partial b} = \frac{1}{h_3}\frac{\partial h_3}{\partial n}\,\hat{b}
\tag{9.65}
$$

$$
\frac{\partial \hat{b}}{\partial s} = \frac{1}{h_1}\frac{\partial h_1}{\partial b}\,\hat{t}, \quad \frac{\partial \hat{b}}{\partial n} = \frac{1}{h_2}\frac{\partial h_2}{\partial b}\,\hat{n}, \quad \frac{\partial \hat{b}}{\partial b} = -\frac{1}{h_3}\frac{\partial h_3}{\partial s}\,\hat{t} - \frac{1}{h_3}\frac{\partial h_3}{\partial n}\,\hat{n}
$$

These relations, unlike Equation (3.43), are not restricted to a two-dimensional or axisymmetric shock. Morse and Feshbach (1953) also provide these orthonormal basis relations. Equation (9.64b) now becomes

$$\vec{V}\cdot\left(\nabla\vec{V}\right)=\left[u\frac{\partial u}{\partial s}+v\frac{\partial u}{\partial n}+v\left(\frac{u}{h_1}\frac{\partial h_1}{\partial n}-\frac{v}{h_2}\frac{\partial h_2}{\partial s}\right)\right]\hat{t}+\left[u\frac{\partial v}{\partial s}+v\frac{\partial v}{\partial n}\right.$$

$$\left.+u\left(\frac{v}{h_2}\frac{\partial h_2}{\partial s}-\frac{u}{h_1}\frac{\partial h_1}{\partial n}\right)\right]\hat{n}+\left(v\frac{\partial\overline{w}}{\partial n}-\frac{u^2}{h_1}\frac{\partial h_1}{\partial b}-\frac{v^2}{h_2}\frac{\partial h_2}{\partial b}\right)\hat{b} \quad (9.64c)$$

Our next task is the evaluation of the h_i derivatives:

$$\frac{1}{h_1}\frac{\partial h_1}{\partial n}=\hat{t}\cdot\frac{\partial\hat{n}}{\partial s}, \quad \frac{1}{h_2}\frac{\partial h_2}{\partial s}=\hat{n}\cdot\frac{\partial\hat{t}}{\partial n}, \quad \frac{1}{h_1}\frac{\partial h_1}{\partial b}=\hat{t}\cdot\frac{\partial\hat{b}}{\partial s}, \quad \frac{1}{h_2}\frac{\partial h_2}{\partial b}=\hat{n}\cdot\frac{\partial\hat{b}}{\partial n} \quad (9.66)$$

that appear in the above equation. As shown by Morse and Feshbach (1953), the curvatures of the s, b, and n coordinates are, respectively,

$$S_a=-\hat{t}\cdot\frac{\partial\hat{n}}{\partial s}=-\frac{1}{h_1}\frac{\partial h_1}{\partial n} \quad (9.67a)$$

$$S_b=-\hat{b}\cdot\frac{\partial\hat{n}}{\partial b}=-\frac{1}{h_3}\frac{\partial h_3}{\partial n} \quad (9.67b)$$

$$\hat{n}\cdot\frac{\partial\hat{t}}{\partial n}=\frac{1}{h_2}\frac{\partial h_2}{\partial s}=n\text{-coordinate curvature} \quad (9.67c)$$

where S_b will appear later in the divergence of the velocity. The n-coordinate curvature, previously denoted as κ_o (see Equation 4.8), must be deleted for a surface evaluation. The remaining two factors in Equation (9.66), $(\partial h_1/\partial b)/h_1$ and $(\partial h_2/\partial b)/h_2$, are next evaluated.

With the use of Equation (G.17a), we have

$$\frac{1}{h_1}\frac{\partial h_1}{\partial b}=\hat{t}\cdot\frac{\partial\hat{b}}{\partial s}=\hat{t}\cdot\frac{\chi}{|\nabla F|}\sum L_j\frac{\partial\hat{b}}{\partial x_j}$$

$$=-\frac{\chi}{|\nabla F|}\sum L_i\,\hat{l}_i\cdot\frac{\chi}{|\nabla F|}\sum L_j\frac{\partial}{\partial x_j}\left(\chi\sum K_k\hat{l}_k\right)$$

$$=-\frac{\chi^2}{|\nabla F|^2}\sum L_i\,\hat{l}_i\cdot\sum L_j\left(\frac{\partial\chi}{\partial x_j}\sum K_k\hat{l}_k+\chi\sum\frac{\partial K_k}{\partial x_j}\hat{l}_k\right)$$

$$=-\frac{\chi^2}{|\nabla F|^2}\sum L_i\left[\left(\sum L_j\frac{\partial\chi}{\partial x_j}\right)\sum K_k\delta_{ik}+\chi\sum_k\sum_j L_j\frac{\partial K_k}{\partial x_j}\delta_{ik}\right]$$

$$=-\frac{\chi^2}{|\nabla F|^2}\left[\left(\sum K_iL_i\right)\sum L_j\frac{\partial\chi}{\partial x_j}+\chi\sum_i L_i\sum_j L_j\frac{\partial K_i}{\partial x_j}\right]$$

$$= -\frac{\chi^3}{|\nabla F|^2} \sum_i L_i \sum_j L_j \frac{\partial K_i}{\partial x_j} \tag{9.68a}$$

since the $K_i L_i$ summation is zero, see Equation (9.13). A similar calculation for the other factor yields

$$\frac{1}{h_2}\frac{\partial h_2}{\partial b} = \hat{n} \cdot \frac{\partial \hat{b}}{\partial n} = -\frac{\chi}{|\nabla F|^2} \sum_i F_{x_i} \sum_j F_{x_j} \frac{\partial K_i}{\partial x_j} \tag{9.68b}$$

where Equation (G.17c) is now used. The state 2 acceleration thus has its final form

$$\vec{V} \cdot (\nabla \vec{V}) = \left(u\frac{\partial u}{\partial s} + v\frac{\partial u}{\partial n} - uv\,S_a \right)\hat{t} + \left(u\frac{\partial v}{\partial s} + v\frac{\partial v}{\partial n} + u^2\,S_a \right)\hat{n}$$

$$+ \left(v\frac{\partial \overline{w}}{\partial n} - \frac{u^2}{h_1}\frac{\partial h_1}{\partial b} - \frac{v^2}{h_2}\frac{\partial h_2}{\partial b} \right)\hat{b} \tag{9.64d}$$

The divergence of the velocity is given by

$$\nabla \cdot \vec{V} = \frac{\partial u}{\partial s} + u\hat{t}\cdot\frac{\partial \hat{t}}{\partial s} + v\hat{t}\cdot\frac{\partial \hat{n}}{\partial s} + \overline{w}\hat{t}\cdot\frac{\partial \hat{b}}{\partial s} + u\hat{n}\frac{\partial \hat{t}}{\partial n} + \frac{\partial v}{\partial n} + v\hat{n}\frac{\partial \hat{n}}{\partial n}$$

$$+ \overline{w}\hat{n}\cdot\frac{\partial \hat{b}}{\partial n} + u\hat{b}\cdot\frac{\partial \hat{t}}{\partial b} + v\hat{b}\cdot\frac{\partial \hat{n}}{\partial b} + \frac{\partial \overline{w}}{\partial b} + \overline{w}\hat{b}\cdot\frac{\partial \hat{b}}{\partial b} \tag{9.69a}$$

Since

$$\hat{t}\cdot\hat{t} = \hat{n}\cdot\hat{n} = \hat{b}\cdot\hat{b} = 1$$

we have

$$\hat{t}\cdot\frac{\partial \hat{t}}{\partial s} = \hat{n}\cdot\frac{\partial \hat{n}}{\partial n} = \hat{b}\cdot\frac{\partial \hat{b}}{\partial b} = 0 \tag{9.70}$$

With these and the foregoing relations, we obtain

$$\nabla \cdot \vec{V} = \frac{\partial u}{\partial s} + \frac{\partial v}{\partial n} - v\,(S_a + S_b) + u\hat{b}\cdot\frac{\partial \hat{t}}{\partial b} \tag{9.69b}$$

The rightmost factor is evaluated in the same way as with Equation (9.68):

$$\frac{1}{h_3}\frac{\partial h_3}{\partial s} = \hat{b}\cdot\frac{\partial \hat{t}}{\partial b} = -\hat{t}\cdot\frac{\partial \hat{b}}{\partial b} = -\frac{\chi^3}{|\nabla F|} \sum_i L_i \sum_j K_j \frac{\partial K_i}{\partial x_j} \tag{9.68c}$$

The divergence of the velocity then is

$$\nabla \cdot \vec{V} = \frac{\partial u}{\partial s} + \frac{\partial v}{\partial n} - v\left(S_a + S_b\right) + \frac{u}{h_3}\frac{\partial h_3}{\partial s} \tag{9.69c}$$

It is worth noting that Equation (9.64d) and Equation (9.69c) are unaltered when the shock is unsteady.

An asterisk is temporarily used to denote a nondimensionial parameter—that is,

$$u^* = \frac{u}{V_1}, \quad v^* = \frac{v}{V_1}, \quad p^* = \frac{p}{p_1}, \quad \rho^* = \frac{\rho}{\rho_1}, \quad \overline{w}^* = \frac{\overline{w}}{V_1}$$

$$\left(\frac{\partial u}{\partial n}\right)^* = \frac{1}{V_1}\frac{\partial u}{\partial n}, \quad \left(\frac{\partial u}{\partial s}\right)^* = \frac{1}{V_1}\frac{\partial u}{\partial s}, \dots$$

This nondimensionalization does not imply that V_1, p_1, or ρ_1 are constants. Note that the $1/V_1$ is outside the $\partial u/\partial n$ derivative, in conformity with Appendix E, and, as a consequence, a $1/(\gamma m)$ factor will appear in the scalar momentum equations. All lengths, including S_a^{-1} and S_b^{-1}, can be dimensional or nondimensional. The final, state 2, nondimensional Euler equations, with the asterisks deleted, are

$$\frac{\partial u}{\partial s} + \frac{\partial v}{\partial n} + \frac{u}{\rho}\frac{\partial \rho}{\partial s} + \frac{v}{\rho}\frac{\partial \rho}{\partial n} - v\left(S_a + S_b\right) + \frac{u}{h_3}\frac{\partial h_3}{\partial s} = 0 \tag{9.71a}$$

$$u\frac{\partial u}{\partial s} + v\frac{\partial u}{\partial n} - uvS_a + \frac{1}{\gamma m \rho}\frac{\partial p}{\partial s} = 0 \tag{9.71b}$$

$$u\frac{\partial v}{\partial s} + v\frac{\partial v}{\partial n} + u^2 S_a + \frac{1}{\gamma m \rho}\frac{\partial p}{\partial n} = 0 \tag{9.71c}$$

$$v\frac{\partial \overline{w}}{\partial n} - \frac{u^2}{h_1}\frac{\partial h_1}{\partial b} - \frac{v^2}{h_2}\frac{\partial h_2}{\partial b} + \frac{1}{\gamma m \rho}\frac{\partial p}{\partial b} = 0 \tag{9.71d}$$

$$u\frac{\partial p}{\partial s} + v\frac{\partial p}{\partial n} - \frac{\gamma p}{\rho}\left(u\frac{\partial \rho}{\partial s} + v\frac{\partial \rho}{\partial n}\right) = 0 \tag{9.71e}$$

In the above, the unknowns are listed in Equation (9.62), parameters, such as u, v, $\partial u/\partial s$, $\partial p/\partial b$, …, are given by the jump and tangential derivative relations, S_a and S_b are provided in Section 9.4, and the three h_i derivatives are given by Equation (9.68). These quantities are fully determined by the Equation (9.1) parameters.

While Equation (9.71b,c,e) are in accord with their Equation (4.10) counterparts, this is not the case for continuity or Equation (9.71d), which has no Equation (4.10) equivalent. When the shock is two-dimensional or axisymmetric,

$$\frac{\partial \bar{w}}{\partial n} = 0, \quad \frac{\partial p}{\partial b} = 0$$

in Equation (9.71d) and one can show that

$$\frac{1}{h_1} \frac{\partial h_1}{\partial b} = \frac{1}{h_2} \frac{\partial h_2}{\partial b} = 0$$

This is done by evaluating the two double sums in Equation (9.68a,b) using Appendix I and

$$F_{x3} = F_{x_2 x_3} = F_{x_3 x_3} = 0, \quad \sigma = 0$$

and appropriate symmetry relations for derivatives of F when $\sigma = 1$. Thus, in the two-dimensional or axisymmetric case, each term in Equation (9.71d) is zero, and $\partial \bar{w}/\partial n$ and Equation (9.71d) are not relevant.

Equation (9.71b) provides $\partial u/\partial n$ directly, while Equation (9.71d) provides $\partial \bar{w}/\partial n$. The other three equations are readily solved for $\partial v/\partial n$, $\partial p/\partial n$, and $\partial \rho/\partial n$. The final result is

$$\frac{\partial u}{\partial n} = \frac{1}{v}\left(-u\frac{\partial u}{\partial s} + uvS_a - \frac{1}{\gamma m \rho}\frac{\partial p}{\partial s}\right) \tag{9.72a}$$

$$\frac{\partial v}{\partial n} = \frac{1}{\dfrac{p}{m\rho} - \gamma v^2}\left(\frac{p}{m\rho}A_a - vA_c + \frac{1}{\gamma m \rho}A_e\right) \tag{9.72b}$$

$$\frac{\partial p}{\partial n} = \frac{1}{\dfrac{p}{m\rho} - v^2}\left(-\gamma p v\, A_a + \gamma p\, A_c - vA_e\right) \tag{9.72c}$$

$$\frac{\partial \rho}{\partial n} = \frac{1}{\dfrac{p}{m\rho} - v^2}\left(-\rho v A_a + \rho A_c - \frac{1}{\gamma m v}A_e\right) \tag{9.72d}$$

$$\frac{\partial \bar{w}}{\partial n} = \frac{1}{v}\left(\frac{u^2}{h_1}\frac{\partial h_1}{\partial b} + \frac{v^2}{h_2}\frac{\partial h_2}{\partial b} - \frac{1}{\gamma m \rho}\frac{\partial p}{\partial b}\right) \tag{9.72e}$$

where

$$A_a = -\frac{\partial u}{\partial s} - \frac{u}{\rho}\frac{\partial p}{\partial s} + v(S_a + S_b) - \frac{u}{h_3}\frac{\partial h_3}{\partial s} \tag{9.73a}$$

$$A_c = -u\,\frac{\partial v}{\partial s} - u^2 S_a \tag{9.73b}$$

$$A_e = -u\,\frac{\partial p}{\partial s} + \frac{\gamma p u}{\rho}\,\frac{\partial \rho}{\partial s} \tag{9.73c}$$

As noted with respect to Equation (8.27), the denominator, $p/(m\rho)-v^2$, in Equation (9.72b,c,d) is zero only when the shock becomes a Mach wave.

Appendices H.5, I.5, and J.5 contain the relevant summaries, although the simplification is quite limited in Appendix I.5. Appendix J.5 contains extra material in view of the complexity of the results for an EP shock.

The $\partial p/\partial n$ derivative, in a tedious effort, has been evaluated for an EP shock with $r = r_2 = r_3$. The result is

$$\frac{\partial p}{\partial n} = \frac{2}{\gamma+1}\,\frac{1}{r\theta^{3/2}XZ}\left[\frac{2}{\gamma+1}\,XYZ(1+\sigma\theta) - \left(\frac{y}{r}\right)^2 g_5 w\right]$$

where g_5 is defined in Appendix E.4 and

$$\theta = 1 + \left(\frac{y}{r}\right)^2,\; y = \begin{cases} x_2, & \sigma = 0 \\ y, & \sigma = 1 \end{cases}$$

Because the shock is two-dimensional or axisymmetric, the derivative should agree with its counterpart in Appendix E.3. Although the two $\partial p/\partial n$ equations do not resemble each other, they are, for both $\sigma = 0$ and $\sigma = 1$, in fact, the same.

An additional check uses an EP shock with

$$\gamma = 1.4,\quad M_1 = 3,\quad w = 4,\quad r_2 = r_3 = 2,\quad \sigma = 0, 1 \tag{9.74}$$

This also is the subject of Problem 26. The data are sufficient to establish the x_i as well as the nondimensional values tabulated in Appendix J.5. For instance, we have

$$m = M_1^2 = 9,\qquad \sin\beta = \left(\frac{w}{m}\right)^{1/2} = \frac{2}{3}$$

$$|\nabla F| = \frac{1}{\sin\beta} = 1.5 = \left[1 + \left(\frac{x_2}{2}\right)^2 + \left(\frac{\sigma x_3}{r_3}\right)^2\right]^{1/2}$$

When $\sigma = 0$, we readily obtain

$$x_2 = 5^{1/2} = 2.236,\quad x_3 = 0$$

When $\sigma = 1$, we have

$$y = \left(x_2^2 + x_3^2\right)^{1/2} = 5^{1/2} = 2.236$$

The coordinates are then given as

$$x_1 = \begin{cases} 1.25, & \sigma = 0 \\ 1.25, & \sigma = 1 \end{cases}, \quad x_2 = \begin{cases} 2.236 \\ 1.581 \end{cases}, \quad x_3 = \begin{cases} 0 \\ 1.581 \end{cases}$$

Problem 26 demonstrates that the s and n derivatives of u, v, p, and ρ check against their Appendix E values. The u, v, p, and ρ values themselves are directly provided by Appendix E. For the Problem 26 data, we obtain

$$\beta = 41.81°, \quad M_2 = 1.816, \quad \frac{\partial p}{\partial \tilde{s}} = -\begin{cases} 3.325 \\ 2.769 \end{cases}$$

and the flow is expansive at state 2. Consequently, there is a Thomas point at a larger β value for both the two-dimensional and axisymmetric shocks.

9.8 APPLICATIONS

A few uses of the theory in this chapter are briefly outlined. The first application discusses the isobaric curve on the shock. The second item provides the angle, λ, between the b and b_{sh} coordinates, where b_{sh} is along the isobaric curve on the shock. The next item evaluates the streamline derivative at state 2. The $\partial p/\partial n$ derivative for a normal EP shock is then derived. The section concludes by establishing the intrinsic coordinate basis at state 2.

As mentioned in Section 9.3, when the shock is three-dimensional, the osculating and flow planes do not coincide and $\tilde{b}$ and $\hat{b}$ are not collinear. Somewhat more general results than those given in Section 9.3 are obtainable. For instance, the isobaric surface condition

$$\left(\frac{\partial p}{\partial b_{sh}}\right)_2 = 0 \tag{9.75}$$

is reexamined, where b_{sh} is a surface coordinate that differs from b. Both b_{sh} and $\tilde{b}$ are also distinct coordinates (as previously noted, $\tilde{b}$ is not in the tangent plane of the shock) that originate at the same state 2 point, and both have a zero pressure gradient. Because $\tilde{b}$ is normal to the osculating plane, b_{sh} cannot also be normal to this plane, otherwise the two coordinates would not be distinct.

The included angle, λ, between the b and b_{sh} surface coordinates is now evaluated. In the shock's tangential plane, we have

$$\frac{\partial p}{\partial b_{sh}} = \frac{\partial p}{\partial b}\cos\lambda + \frac{\partial p}{\partial s}\sin\lambda = 0 \qquad (9.76)$$

This yields

$$\tan\lambda = -\frac{\dfrac{\partial p}{\partial b}}{\dfrac{\partial p}{\partial s}} \qquad (9.77a)$$

where the s and b derivatives are given by Equations (H.17) and (H.18). When the upstream flow is uniform, Appendix I simplifies this relation to

$$\tan\lambda = |\nabla F|\,\frac{\sum K_i\,\beta_{x_i}}{\sum L_i\,\beta_{x_i}} \qquad (9.77b)$$

Because $\tilde{s}$ is in the flow plane, the $\partial()/\partial\tilde{s}$ derivative only depends on the $\partial()/\partial s$ and $\partial()/\partial n$ derivatives. In other words, the direction cosine between $\hat{b}$ and $\hat{\tilde{s}}$ is zero (see Equation 9.15b). Hence, the derivative, $\partial()/\partial\tilde{s}$, is still given by Equation (5.10a).

General results for a normal derivative, at a location where the shock is normal to the upstream flow, are not as straightforward as in Section 5.1. These equations are indeterminate. For instance, we start with the normal shock condition

$$\frac{\bar{V}_1}{V_1} = \hat{n}$$

which yields

$$\sin\beta = 1, \quad \cos\beta = 0$$

as expected. But now χ, which appears in many equations, is infinite, while the K_j and L_j, for all j, are zero.

To simplify and focus the discussion, the $\partial p/\partial n$ derivative is evaluated at the nose of an EP shock, with $r_2 \neq r_3$. This location is at the origin of the x_i-coordinate system. To evaluate the indeterminacies, a plane is introduced that passes through the x_1-coordinate:

$$x_3 = x_2\tan\alpha$$

where α is a given angle. The indeterminacies are evaluated, as the origin is approached, along a curve given by the intersection of the plane with the shock. It is convenient to introduce the following constants:

$$c_1^2 = \frac{1}{r_2} + \frac{\tan^2\alpha}{r_3} \qquad (9.79a)$$

$$c_2^2 = \frac{1}{r_2^2} + \sigma \left(\frac{\tan\alpha}{r_3} \right)^2 \qquad (9.79b)$$

$$c_3^2 = \frac{1}{r_2^3} + \sigma \frac{\tan^2\alpha}{r_3^3} \qquad (9.79c)$$

In the subsequent listing, only the leading-order term, when $x_2 \to 0$, is provided.

$$\psi = c_2^2 \, x_2^2 \qquad (9.80a)$$

$$\sin\beta = 1, \quad \cos\beta = c_2 x_2 \qquad (9.80b)$$

$$\frac{x_2^2}{r_2} + \frac{x_3^2}{r_3} = c_1^2 x_2^2 \qquad (9.80c)$$

$$\frac{x_2^2}{r_2^2} + \sigma \frac{x_3^2}{r_3^2} = c_2^2 \, x_2^2 \qquad (9.80d)$$

$$\frac{x_2^2}{r_2^3} + \sigma \frac{x_3^2}{r_3^3} = c_3^2 \, x_2^2 \qquad (9.80e)$$

$$u = c_2 x_2, \quad v = \frac{2}{\gamma+1} \frac{X}{m}, \quad p = \frac{2}{\gamma+1} Y, \quad \rho = \frac{\gamma+1}{2} \frac{m}{X} \qquad (9.80f)$$

$$\frac{1}{h_1} \frac{\partial h_1}{\partial b} = \frac{\sigma}{c_2^3 r_3 r_2} \left(\frac{1}{r_3} - \frac{1}{r_2} \right) \frac{\tan\alpha}{x_2}, \quad \frac{1}{h_2} \frac{\partial h_2}{\partial b} = 0, \quad \frac{1}{h_3} \frac{\partial h_3}{\partial s} = \frac{c_1^2}{r_2 r_3 c_2^3} \frac{\sigma}{x_2} \qquad (9.80g)$$

$$S_a = \left(\frac{c_3}{c_2} \right)^2, \quad S_b = \frac{\sigma}{r_2 r_3} \left(\frac{c_1}{c_2} \right)^2 \qquad (9.80h)$$

$$\frac{\partial u}{\partial s} = \left(\frac{c_3}{c_2} \right)^2, \quad \frac{\partial v}{\partial s} = \frac{\partial p}{\partial s} = \frac{\partial \rho}{\partial s} = 0 \qquad (9.80i)$$

$$A_a = -\frac{2}{\gamma+1} \frac{Z}{m} \left[\left(\frac{c_3}{c_2} \right)^2 + \frac{\sigma}{r_2 r_3} \left(\frac{c_1}{c_2} \right)^2 \right], \quad A_c = A_e = 0 \qquad (9.80j)$$

$$\frac{p}{m\rho} - v^2 = \frac{2}{\gamma+1} \frac{XZ}{mw} \qquad (9.80k)$$

Note that A_a may be written in terms of the S_a, S_b curvatures, and the s derivatives of v, p, and ρ are zero as expected from symmetry considerations at a location where the shock is normal to the upstream flow.

With the foregoing, the desired pressure gradient, given by Equation (J.43), is

$$\frac{\partial p}{\partial n} = \frac{4\gamma}{(\gamma+1)^2} Y \left[\left(\frac{c_3}{c_2}\right)^2 + \frac{\sigma}{r_2 r_3} \left(\frac{c_1}{c_2}\right)^2 \right] = \frac{4\gamma}{(\gamma+1)^2} Y (S_a + S_b) \tag{9.81}$$

This readily reduces to Equation (5.5b) when the shock is two-dimensional or axisymmetric. In the axisymmetric case, set $r_2 = r_3 = R_s$. This gradient not only depends on γ, M_1, and σ, but also on the sum of the curvatures.

A right-handed, orthonormal, intrinsic coordinate basis, $\hat{\tilde{s}}, \hat{\tilde{b}}, \hat{\tilde{n}}$, at state 2, is now established. Since $\hat{\tilde{s}}$, given by Equation (9.15b), is in terms of $\hat{t}$ and $\hat{n}$, it is analytically convenient to continue to use the $\hat{t}, \hat{n}, \hat{b}$ basis. We thus write

$$\hat{\tilde{b}} = a_t \hat{t} + a_n \hat{n} + a_b \hat{b} \tag{9.82a}$$

where

$$a_t^2 + a_n^2 + a_b^2 = 1 \tag{9.82b}$$

The equation for $\hat{\tilde{n}}$ is then

$$\hat{\tilde{n}} = \hat{\tilde{s}} \times \hat{\tilde{b}} = \frac{1}{B^{1/2}} \left[a_b \hat{t} - A a_b \hat{n} + (A a_n - a_t) \hat{b} \right] \tag{9.83a}$$

A second relation for the a's stems from

$$\hat{\tilde{b}} \cdot \hat{\tilde{s}} = 0$$

or

$$a_n = - a_t A \tag{9.84}$$

As has been established, the pressure gradient is zero along the coordinate tangent to $\hat{\tilde{b}}$—that is,

$$\left(\frac{\partial p}{\partial \tilde{b}}\right)_2 = \hat{\tilde{b}} \cdot \nabla p = 0$$

which becomes

$$a_t \frac{\partial p}{\partial s} + a_n \frac{\partial p}{\partial n} + a_b \frac{\partial p}{\partial b} = 0 \tag{9.85}$$

where equations for the pressure derivatives are provided in Appendices H, I, and J.

Equations (9.82b), (9.84), and (9.85) yield the desired solution:

$$a_t = \pm \frac{1}{D} \frac{\partial p}{\partial b}, \quad a_b = \mp \frac{1}{D} \left(\frac{\partial p}{\partial s} - A \frac{\partial p}{\partial n} \right), \quad a_n = \mp \frac{A}{D} \frac{\partial p}{\partial b} \qquad (9.86)$$

where

$$D = \left[\left(\frac{\partial p}{\partial s} - A \frac{\partial p}{\partial n} \right)^2 + B \left(\frac{\partial p}{\partial b} \right)^2 \right]^{1/2} \qquad (9.87)$$

It is expected that $\hat{b} \cdot \hat{b} > 0$, or $a_b > 0$. With Equations (5.10b) and (5.22), a_b can be written as

$$a_b = \pm \frac{B^{1/2}}{D} \left(\rho V^2 \right)_2 \frac{\partial \theta}{\partial \tilde{s}} \qquad (9.88)$$

From the Crocco point discussion in Section 5.6, it is evident that $(\theta / \partial \tilde{s})$ can be positive or negative. Hence, both signs are required in Equation (9.86) in order that $a_b > 0$.

By way of summary, $\hat{s}$ is given by Equation (9.15b), $\hat{b}$ is given by Equations (9.82a) and (9.86), and $\hat{n}$ by Equation (9.83a). In view of Equations (9.84) and (9.86), the $\hat{n}$ equation can be written as

$$\hat{n} = \frac{1}{B^{1/2}D} \left[\left(\frac{\partial p}{\partial s} + \frac{\partial p}{\partial n} \right) \left(\mp \hat{\imath} \pm A\hat{n} \right) \mp B \frac{\partial p}{\partial b} \hat{b} \right] \qquad (9.83b)$$

As an illustration, the intrinsic coordinate basis is evaluated for an EP shock whose data are given by Equation (9.74), and which is the subject of Problem 26. Appendix J is utilized to obtain

$$A_a = \begin{cases} 0.07545, & \sigma = 0 \\ -0.06344, & \sigma = 1 \end{cases}, \quad A_c = -0.08573, \quad A_e = 0.02881$$

$$\frac{\partial p}{\partial s} = -1.546, \quad \frac{\partial p}{\partial b} = 0, \quad \frac{\partial p}{\partial n} = \begin{cases} -5.848 \\ -4.098 \end{cases} \qquad (9.89)$$

$$D = \begin{cases} 15.89 \\ 10.67 \end{cases}$$

with the result

$$a_t = 0, \quad a_n = 0, \quad a_b = 1 \qquad (9.90)$$

This simple result could have been anticipated, because the shock is two-dimensional or axisymmetric. This is evident from $r_2 = r_3$, or $r_3 = 0$, and $\partial p/\partial b$ equaling zero. The intrinsic coordinate basis therefore is

$$\hat{s} = \frac{1}{B^{1/2}}\left(A\hat{t} + \hat{n}\right) \tag{9.91a}$$

$$\hat{b} = \hat{b} \tag{9.91b}$$

$$\hat{n} = \frac{1}{B^{1/2}}\left(\hat{t} - A\hat{n}\right) \tag{9.91c}$$

The $\hat{s}$ equation is Equation (9.15b), and the $\hat{n}$ equation is also derived in Problem 9, where it is denoted as $\hat{e}_{no}$.

9.9 UNSTEADY, NORMAL DERIVATIVE FORMULATION

The approach in Section 9.7 is extended to an unsteady shock. Much of the earlier material still applies. This includes Appendix E.1 for jump conditions, and Section 9.2, where the $\hat{t},\hat{n},\hat{b}$ basis is obtained. Equations (9.21a) through (9.24b) and Equation (9.26) for the S_a, S_b curvatures, as well as Equations (9.59) through (9.61), for $\bar{V}, \nabla$, and $\bar{w}$, hold. Equations (9.64d) and (9.69c) for $\bar{V}\cdot(\nabla\bar{V})$ and $\nabla\cdot\bar{V}$, respectively, along with Equation (9.68), for the scale factor derivatives, also hold. On the other hand, the vorticity analysis in Section 9.5 does not apply. This is because the acceleration term, $(\partial\bar{V}/\partial t)_2$, in Crocco's equation that is used in the vorticity derivation, was not included. (This derivative is evaluated shortly.)

One change, however, is that the $v_{1,i}$ item in Equation (9.1) is relabeled as $v'_{1,i}$, where

$$\bar{V}_1' = \sum v'_{1,i}\hat{l}_i \tag{9.92}$$

As indicated in Figure 9.2, $\bar{V}_1'$, which is not necessarily uniform, is the upstream velocity. The local velocity, however, is

$$\bar{V}_1 = \bar{V}_1' - \bar{V}_R = \sum v_{1,i}\hat{l}_i \tag{9.93}$$

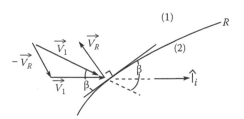

FIGURE 9.2 Sketch of the $\bar{V}_1$, $\bar{V}_1'$, and $\bar{V}_R$ velocities.

Where the shock's velocity, given by Equation (2.5), is

$$\vec{V}_R = V_R \hat{n} = \left(-\frac{1}{|\partial F|} \frac{\partial F}{\partial t}\right) \frac{\partial F}{|\nabla F|} = -\frac{F_t}{|\nabla F|^2} \sum F_{x_i} \hat{l}_i \qquad (9.94a)$$

As noted in Figure 9.2, β is measured from $\vec{V}_1$, not $\vec{V}_1'$. The magnitude of $\vec{V}_R$ is readily obtained

$$V_R = \frac{F_t}{|\nabla F|} \qquad (9.94b)$$

Combining the above yields

$$v_{1,i} = v_{1,i}' + \frac{F_t F_{x_i}}{|\nabla F|^2} \qquad (9.95a)$$

and

$$V_1 = \left(\sum v_{1,i}'^2 + \frac{2F_t}{|\nabla F|^2} \sum F_{x_i} v_{1,i}' + \frac{F_t^2}{|\nabla F|^2}\right)^{1/2} \qquad (9.95b)$$

With $v_{1,i}'$ changed to $v_{1,i}$, the equations in Section 9.2 are unaltered.

The unsteady Euler equations are

$$\frac{\partial \rho}{\partial t} + \rho \nabla \cdot \vec{V} + \vec{V} \cdot \nabla \rho = 0 \qquad (9.96a)$$

$$\frac{\partial \vec{V}}{\partial t} + \vec{V} \cdot \nabla \vec{V} + \frac{1}{\rho} \nabla p = 0 \qquad (9.96b)$$

$$\frac{\partial p}{\partial t} - \gamma \frac{p}{\rho} \frac{\partial \rho}{\partial t} + u \frac{\partial p}{\partial s} + v \frac{\partial p}{\partial n} - \gamma \frac{p}{\rho}\left(u \frac{\partial \rho}{\partial s} + v \frac{\partial \rho}{\partial n}\right) = 0 \qquad (9.96c)$$

The normal derivative solution, as before, depends on state 2 parameters, such as p_2, and state 2 tangential derivatives, both with respect to s and b. The normal derivates also depend on

$$\left(\frac{\partial p}{\partial t}\right)_2, \quad \left(\frac{\partial \rho}{\partial t}\right)_2, \quad \left(\frac{\partial \vec{V}}{\partial t}\right)_2 \qquad (9.97)$$

where $\left(\partial \vec{V}/\partial t\right)_2$ must be written in terms of the $\hat{t}, \hat{n}, \hat{b}$ basis. All these parameters are to be evaluated in terms of the Equation (9.1) items, but with the $v_{1,i}'$ replacement.

Evaluation of $\left(\partial \vec{V}/\partial t\right)_2$ starts with

$$\vec{V}_2 = \sum v_{2,i} \hat{l}_i \qquad (9.98a)$$

and

$$\left(\frac{\partial \vec{V}}{\partial t}\right)_2 = \sum v_{2,it} \hat{l}_i \qquad (9.99a)$$

By inverting the $\hat{t}, \hat{n}, \hat{b}$ equations in Section 9.2, we obtain

$$\hat{l}_i = \frac{\chi}{|\nabla F|} L_i \hat{t} + \frac{1}{|\nabla F|} F_{x_i} \hat{n} - \chi K_i \hat{b}, \qquad i = 1, 2, 3 \qquad (9.100)$$

with the result

$$\left(\frac{\partial \vec{V}}{\partial t}\right)_2 = \frac{\chi}{|\nabla F|} \hat{t} \sum L_i v_{2,it} + \frac{\hat{n}}{|\nabla F|} \sum F_{x_i} v_{2,it} - \chi \hat{b} \sum K_i v_{2,it} \qquad (9.99b)$$

From Equations (9.14) and (9.15b), $\vec{V}_2$ can also be written as

$$\vec{V}_2 = \frac{V_2}{B^{1/2}} (\hat{n} + A\hat{t}) \qquad (9.98b)$$

where (see Appendix E.1)

$$V_2 = \frac{2}{\gamma+1} V_1 \frac{XB^{1/2} \sin\beta}{w} \qquad (9.101)$$

Hence, $v_{2,i}$ is given by

$$v_{2,i} = \frac{V_2}{|\nabla F| B^{1/2}} \left(F_{x_i} + \chi A L_i\right), \qquad i = 1, 2, 3 \qquad (9.102)$$

Its time derivative is written as

$$v_{2,it} = v_{2,i}(J + H_i), \qquad i = 1, 2, 3 \qquad (9.103)$$

where J, which does not depend on the i suffix, and H_i are given in Appendix L by Equations (L.28) and (L.29), respectively. The time derivatives of χ, A, and L_i that appear in H_i are listed in Appendix L.

With the above and Equations (9.64d), (9.69c), and (9.68), the Euler equations can be written as:

$$\frac{\partial v}{\partial n} + \frac{v}{\rho} \frac{\partial \rho}{\rho n} = A_1 \qquad (9.104a)$$

$$v \frac{\partial u}{\partial n} = A_2 \qquad (9.104b)$$

$$v \frac{\partial v}{\partial n} + \frac{1}{\rho} \frac{\partial p}{\partial n} = A_3 \qquad (9.104c)$$

$$v\frac{\partial \overline{w}}{\partial n} = A_4 \tag{9.104d}$$

$$\frac{\partial p}{\partial n} - \gamma \frac{p}{\rho}\frac{\partial \rho}{\partial n} = A_5 \tag{9.104e}$$

where

$$A_1 = -\frac{\partial u}{\partial s} + v(S_a + S_b) + \frac{\chi^3 u}{|\nabla F|}\sum_{i,j} L_j K_j \frac{\partial K_i}{\partial x_j} - \frac{u}{\rho}\frac{\partial \rho}{\partial s} - \frac{1}{\rho}\frac{\partial \rho}{\partial t} \tag{9.105a}$$

$$A_2 = -u\frac{\partial u}{\partial s} + uvS_a - \frac{1}{\rho}\frac{\partial p}{\partial s} - \frac{\chi}{|\nabla F|}\sum L_i v_{2,it} \tag{9.105b}$$

$$A_3 = -u\frac{\partial v}{\partial s} - u^2 S_a - \frac{1}{|\nabla F|}\sum F_{x_i} v_{2,it} \tag{9.105c}$$

$$A_4 = -\frac{\chi^3 u^2}{|\nabla F|^2}\sum_{i,j} L_i L_j \frac{\partial K_i}{\partial x_j} - \frac{\chi v^2}{|\nabla F|^2}\sum_{i,j} F_{x_i} F_{x_j} \frac{\partial K_i}{\partial x_j} - \frac{1}{\rho}\frac{\partial p}{\partial b} + \chi\sum K_i v_{2,it} \tag{9.105d}$$

$$A_5 = -\frac{u}{v}\frac{\partial p}{\partial s} + \gamma\frac{p}{\rho}\frac{u}{v}\frac{\partial \rho}{\partial s} - \frac{1}{v}\frac{\partial p}{\partial t} + \gamma\frac{p}{\rho v}\frac{\partial \rho}{\partial t} \tag{9.105e}$$

Two of the equations have a ready solution,

$$\frac{\partial u}{\partial n} = \frac{1}{v} A_2 \tag{9.106a}$$

$$\frac{\partial \overline{w}}{\partial n} = \frac{1}{v} A_4 \tag{9.106b}$$

The remaining three equations are easily solved, with the result

$$\frac{\partial v}{\partial n} = \frac{1}{\frac{\rho v^2}{\gamma p} - 1}\left(-A_1 + \frac{\rho v}{\gamma p} A_3 - \frac{v}{\gamma p} A_5\right) \tag{9.106c}$$

$$\frac{\partial p}{\partial n} = \frac{\rho}{\frac{\rho v^2}{\gamma p} - 1}\left(vA_1 - A_3 + \frac{v^2}{\gamma p} A_5\right) \tag{9.106d}$$

$$\frac{\partial \rho}{\partial n} = \frac{\frac{\rho}{\gamma p}}{\frac{\rho v^2}{\gamma p} - 1}\left(\rho v A_1 - \rho A_3 + A_5\right) \tag{9.106e}$$

These equations agree with the steady results in Section 9.7, when the unsteady terms are deleted.

The tangential derivatives, as before, are based on Appendix E.1. Appendix H.4 is directly applicable for the s and b derivatives. When the shock's unsteady, $\vec{V}_1$ and its components are not constant, and Appendix H.4, which does not assume a uniform upstream flow, is thus applicable. On the other hand, if the shock is propagating into a uniform flow, Appendix I.4 cannot be used, since $\vec{V}_1$ is not constant.

Formulas for the unsteady, tangential, and normal derivatives at state 2 are provided by Appendix L, where the tangential derivatives are provided by Appendix H.4 and the normal derivatives also require Equations (9.105) and (9.106).

9.10   SINGLE MACH REFLECTION

The tangential and normal derivatives are evaluated just downstream of the reflected shock in a single Mach reflection (SMR) pattern (see Figure 9.3). The figure is a side-view sketch of the flow pattern shortly after a planar incident shock encounters a straight ramp inside a shock tube with a rectangular cross-section. The features R, I, M, SS, and T are the reflected shock, incident shock, Mach stem, slipstream, and triple point, respectively. The reflected shock may or may not have a maximum x_2 value. The figure shows a maximum at x_{1c}, x_{2c}, since the experimental reflected shock to be used has a maximum. The origin is at the leading edge of the ramp and x_1 is aligned with the velocity of the incident shock.

Detailed data for the SMR pattern comes from Ben-Dor and Glass (1978). Yi (1999) and Yi and Emanuel (2000) provide analytical/computational results for the vorticity just downstream of the reflected shock, based on data in Figures (4c) and (4d) of Ben-Dor and Glass (1978). Emanuel and Yi (2000) is another unsteady shock analysis, but does not utilize Ben-Dor and Glass (1978). Both Emanuel and Yi (2000) and Yi and Emanuel (2000) are based on Yi's MS thesis, from which Figure 9.3 is taken.

For brevity, our interest in SMR is limited to providing the unsteady, tangential, and normal derivatives for the pressure and density just downstream of the reflected

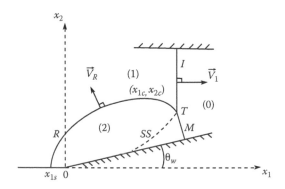

FIGURE 9.3   Schematic of a single-Mach reflection configuration.

shock. The flow field downstream of the incident shock is uniform, and the reflected shock propagates into this flow.

At a given shock point, state 1 is associated with the upstream velocity, $\vec{V}_1$, in contrast to $\vec{V}_1'$, which is constant and is imposed by the incident shock. The shock wave angle, β (see Figure 9.2), is measured relative to $\vec{V}_1$ not $\vec{V}_1'$, and $\left(\partial\vec{V}_1'/\partial t\right)$ is zero, but $\partial\vec{V}_1'/\partial t$ is not. The flow is two-dimensional, which later provides some simplification.

Before embarking on numerical results, the shock shape and its required derivatives are established. This is followed by a discussion of the flow associated with the reflected shock.

As shown in Yi and Emanuel (2000), the reflected shock is approximately modeled as an ellipse, where x_{2c} is the maximum value on the shock and x_{1s} is where it is normal to the wall (see Figure 9.3). For an SMR flow, there is no characteristic length in the initial data, which would not be the case, for example, if the ramp's surface is curved. With a uniform upstream flow and no initial data characteristic length, the reflected shock is referred to as pseudostationary (Glass and Sislian 1994, p. 184). In this circumstance, the reflected shock's shape is invariant with time when measured from the start of the interaction. In other words, the reflected shock at time $2t$ looks just like it does at time t, although doubly magnified. The shock's shape thus is (Yi and Emanuel, 2000)

$$\left(\frac{x_1 - x_{1c}}{x_{1s} - x_{1c}}\right)^2 + \left(\frac{x_2}{x_{2c}}\right)^2 = 1 \tag{9.107}$$

where

$$x_{1s} = -c_1 a_0 t, \quad x_{1c} = c_2 a_0 t, \quad x_{2c} = c_3 a_0 t \tag{9.108}$$

The c_i are positive, nondimensional constants, and a_0 is the speed of sound in the quiescent upstream region. It is convenient to write the shock shape as (Yi, 1999; Yi and Emanuel, 2000)

$$F(x_1, x_2, t) = f(x_1, t) - x_2 = 0 \tag{9.109a}$$

where

$$x_2 = f = \frac{c_3}{c_1 + c_2} A^{1/2} B^{1/2} \tag{9.109b}$$

$$A = (c_1 + 2c_2)a_0 t - x_1 \tag{9.109c}$$

$$B = c_1 a_0 t + x_1 \tag{9.109d}$$

$$C = c_2 a_0 t - x_1 \tag{9.109e}$$

At the wall, where $x_{2s} = 0$, B is zero and x_{1s} equals $-c_1 a_0 t$.

The various nonzero derivatives of F, required shortly, are given by:

$$F_{x_1} = \frac{C x_2}{AB} \tag{9.110a}$$

$$F_{x_2} = -1 \tag{9.110b}$$

$$F_t = \frac{a_0 x_2}{AB}\left[c_1(c_1 + 2c_2)a_0 t + c_2 x_1\right] \tag{9.110c}$$

$$F_{x_1 x_1} = \frac{F_{x_1}^2}{x_2} + \frac{F_{x_1}}{ABC}(-AB + BC - CA) \tag{9.110d}$$

$$F_{x_1 t} = F_{t x_1} = \frac{F_{x_1} F_t}{x_2} + \frac{a_0 F_{x_1}}{ABC}\left[c_1 AB - (c_1 + 2c_2)BC - c_1 AC\right] \tag{9.110e}$$

$$F_{tt} = \frac{a_0}{AB}\left\{\left[c_1(c_{1+}2c_2)a_0 t + c_2 x_1\right]F_t + c_1(c_{1+}2c_2)a_0 x_2 \right.$$

$$\left. -a_0 x_2\left[c_1(c_1 + 2c_2)a_0 t + c_2 x_1\right]\left[\frac{c_1 + 2c_2}{A} + \frac{c_1}{B}\right]\right\} \tag{9.110f}$$

In obtaining F_{tt}, note that $\partial x_2/\partial t$ equals $\partial f/\partial t$. Although x_2, given by Equation (9.109b), appears in several of these equations, they only depend on x_1 and t_1. As a consequence, parameters such as β_{x_2} are zero. That is, the formula for β will depend only on x_1 and t.

At this time, it is convenient to provide the quiescent gas conditions in the shock tube (Ben-Dor and Glass, 1978) and shock-shape data (Yi, 1999; and Yi and Emanuel, 2000):

$$\gamma = 1.4, \quad R = 296.95\frac{J}{kg - K'} \quad M_I' = 2.01$$

$$p_0 = 6.665 \times 10^3 \, Pa, \quad \rho_0 = 7.567 \times 10^{-2} \, kg/m^3,$$

$$T_0 = 296.6 \, K, \quad a_0 = 351.1 \, m/s \tag{9.111}$$

$$t = 4.817 \times 10^{-5} \, s, \quad c_1 = 0.1898, \quad c_2 = 1.537, \quad c_3 = 1.496 \tag{9.112}$$

The given time is from Ben-Dor and Glass (1978). It is the time when they provide, in Figure 4(c), constant density curves downstream of the Mach disk and reflected shock based on a Mach–Zehnder interferogram. With the foregoing, the sole independent variable is x_1, where $x_{1s} \leq x_1 \leq x_{1T}$. In addition, the ramp angle is 26.56 degrees. Note that the incident shock Mach number, M_I' is modest, which results in a weak reflected shock.

The change in Mach number across an unsteady, normal shock is

$$M_1' = \frac{M_I^2 - 1}{\left(1 + \frac{\gamma - 1}{2}M_I^2\right)^{1/2}\left(\gamma M_I^2 - \frac{\gamma - 1}{2}\right)^{1/2}} = 0.9679 \tag{9.113}$$

The corresponding downstream velocity is

$$\vec{V}_1' = a_1 M_1' \hat{l}_1 = 457.2 \times 0.9679\, \hat{l}_1 = 442.5\, \hat{l}_1,\, m/s \tag{9.114}$$

where the speed of sound is given by

$$a_1 = \left(\frac{T_1}{T_0}\right)^{1/2} a_0 \tag{9.115}$$

From Equation (L.3) in Appendix L, we have

$$v_{1,1} = a_1 M_1' + \frac{F_t F_{x_1}}{|\nabla F|^2}, \quad v_{1,2} = -\frac{F_t}{|\nabla F|^2}, \quad v_{1,3} = 0 \tag{9.116a}$$

where $a_1 M_1' (= V_1')$ is a constant and

$$V_1 = \left(v_{1,1}^2 + v_{1,2}^2\right)^{1/2} \tag{9.116b}$$

$$M_1 = \frac{V_1}{a_1} \tag{9.116c}$$

Also note that

$$p_{1t} = \rho_{1t} = 0 \tag{9.116d}$$

and

$$\frac{M_{1t}}{M_1} = \frac{1}{V_1^2} \sum v_{1,i} v_{1,it} \tag{9.116e}$$

where

$$v_{1,1t} = \frac{F_{x_1} F_{tt}}{|\nabla F|^2} + \frac{F_t F_{x_1t}}{|\nabla F|^4}\left(1 - F_{x_1}^2\right) \tag{9.116f}$$

$$v_{1,2t} = -\frac{F_{tt}}{|\nabla F|^2} + \frac{2}{|\nabla F|^4} F_{x_1} F_t F_{x_1t} \tag{9.116g}$$

Finally, we have

$$\beta = \sin^{-1}\left(\frac{\sum v_{1,i} F_{x_i}}{V_1 |\nabla F|}\right) = \sin^{-1}\left(\frac{v_{1,1} F_{x_1} - v_{1,2}}{V_1 |\nabla F|}\right) \tag{9.116h}$$

where

$$|\nabla F| = \left(\sum F_{x_i}^2\right)^{1/2} = \left(1 + F_{x_1}^2\right)^{1/2} \tag{9.116i}$$

Note that M_1 and β only depend on fixed quiescent gas parameters, M_i', and F and its derivatives.

Consider a ray through the origin that intersects the reflected shock. This ray is written as

$$x_i = b_i t, \qquad i = 1, 2 \tag{9.117}$$

where b_1 is a negative constant when the ray is in the second quadrant, and b_2 is a positive constant. From the equations in Equation (9.109), we obtain

$$b_2 = \frac{c_3}{c_1 + c_2} \left\{ \left[(c_1 + 2c_2) a_0 - b_1 \right] (c_1 a_0 + b_1) \right\}^{1/2} \tag{9.118}$$

Similarly, F_{x_1} and F_t are constant along the ray, for example,

$$F_{x_1} = \frac{Cx_2}{AB} = \frac{(c_2 a_0 - b_1) t b_2 t}{\left[(c_1 + 2c_2) a_0 - b_1 \right] (c_1 a_0 + b_1) t^2}$$

$$= \frac{(c_2 a_0 - b_1) b_2}{\left[(c_1 + 2c_2) a_0 - b \right] (c_1 a_0 + b_1)} \tag{9.119}$$

As a consequence, β and M_1 are also constant along a ray. The β result is necessary if the reflected shock is to retain its shape with time. Thus, parameters, such as p_2 and ρ_2, are also constant along a ray. Second-order F derivatives, however, are not constant along a ray. Hence, constancy along a ray does not extend to derivative quantities. For instance, β_t, which appears in p_{2t} and ρ_{2t}, depends on $F_{x_1 t}$, which approaches infinity as $t \to 0$.

The most obvious simplification when the flow is two-dimensional is

$$\hat{b} = -\hat{1}_3$$

This results in

$$K_1 = K_2 = 0, \quad K_3 = \frac{1}{\chi} = V_1' = constant$$

$$L_1 = K_3, \quad L_2 = K_3 F_{x_1}, \quad L_3 = 0$$

$$L_{1t} = 0, \quad L_{2t} = K_3 F_{x_1 t}, \quad L_{3t} = 0$$

Table 9.1 shows results along the shock, where case 1 is very close to the attachment point. At this point, β is 89.42 degrees. The computer calculation could not be extended to the wall because all F derivatives become infinite at the wall. Case 6 is

TABLE 9.1

Reflected Shock Results ($p_1 = 3.030 \times 10^4$ Pa, $\rho_1 = 0.2029$ kg/m^3, Length in Meters, Velocity in Meters/Seconds)

Case	x_1	M_1	$M_1 \sin\beta$	V_R	$\dfrac{p_2}{p_1}$	$\dfrac{\rho_2}{\rho_1}$	$\dfrac{p_{2l}}{p_1}$	$\dfrac{\rho_{2l}}{\rho_1}$	$\dfrac{p_{2s}}{p_1}$	$\dfrac{\rho_{2s}}{\rho_1}$	$\dfrac{p_{2n}}{p_1}$	$\dfrac{\rho_{2n}}{\rho_1}$
1	−3.208−3*	1.114	1.114	66.66	1.280	1.193	−937.7	−619.2	−0.1638	−0.1082	−57.41	−38.21
2	2.1−3	1.252	1.094	156.4	1.229	1.158	254.2	170.2	−3.673	−2.460	−195.5	−131.7
3	4.76−3	1.302	1.089	198.9	1.218	1.151	261.4	180.6	−1.950	−1.310	−228.4	−154.2
4	1.007−2	1.378	1.090	280.9	1.218	1.151	−563.4	−378.5	2.348	1.577	−243.3	−164.2
5	1.803−2	1.454	1.109	401.4	1.268	1.184	−3607.0	−2390.0	9.358	6.202	−203.8	−136.0
6	2.6−2	1.502	1.149	525.2	1.373	1.253	−9117.0	−5870.0	16.89	10.88	−159.4	−103.9
7	3.4−2	1.532	1.211	659.5	1.543	1.360	−1.844+4	−1.135+4	25.37	15.61	−132.3	−83.58

*Read −3.208−3 as −3.208×10^{-3}.

where the shock's x_2 value has a maximum and case 7 is close to the triple point. An algorithm, based on Appendices H.4 and L, was developed for the reflected shock and ably coded by Dr. Hekiri.

While M_1 gradually increases with arc length along the shock, its normal component, $M_1 \sin\beta$, has a shallow minimum near case 3. Overall, $M_1 \sin\beta$ is nearly constant, and its value indicates the relative weakness of the shock. As expected, both p_2/p_1 and ρ_2/ρ_1 have a similar minimum. Because of this minimum, both p_{2s}/p_1 and ρ_{2s}/ρ_1 are negative at first, but become positive after the minimum.

For a steady, convex shock with a uniform freestream, $M_1 \sin\beta$ steadily decreases. This does not happen here, because the shock speed, V_R, steadily and rapidly increases. This results in the roughly constant shock strength.

Surprisingly, the pressure and density time derivatives go through two zeros, and their magnitude dwarfs that of the other derivatives as the triple point is approached. The magnitude of the various derivatives appears large because seconds and meters are the units rather than ms and mm. As noted, the tangential derivatives have one zero, while the normal derivatives do not.

The negative normal density derivative is in accord with the blast wave theory, in which the density derivative, just downstream of a shock, even when it is weak, is negative. The normal density derivative, which is evident in Figure 4c of Ben-Dor and Glass (1978), however, is positive. Part of the discrepancy may be due to the assumption of an elliptical shape for the shock. For instance, at the attachment point, Ben-Dor and Glass in Figure 4c, show a shock inclination angle, relative to the wall, of about 77 degrees. This is due to a shock interaction with the residual boundary layer from the incident shock. The elliptic shock approximation is thus not reliable in this region.

Appendix A: Selective Nomenclature

a: Speed of sound

A: $\dfrac{\gamma+1}{2}\dfrac{m\,\sin\beta\,\cos\beta}{X}$; cross-sectional area

B: $1 + A^2$

C$_-$: Right-running characteristic wave

C$_+$: Left-running characteristic wave

F: $F = 0$ is the shock's shape

h: Enthalpy

h_i: Scale factors

J: Jacobian

K_j: Equation (9.8)

L_j: Equation (9.11a)

m: M_1^2

$\dot{m}$: Mass flow rate

M: Mach number

s,n,b: Arc length coordinates tangential to $\hat{t}, \hat{n}, \hat{b}$, respectively

$\hat{t}, \hat{n}, \hat{b}$: Right-handed, orthonormal shock basis; $\hat{n}$ normal to the shock in the downstream direction, $\hat{b}$ binormal basis, $\hat{t}$ basis vector tangent to the shock in the flow plane

p: Pressure

R: Gas constant; radius

$\vec{r}$: Position vector

S: Shock surface curvature; entropy

T: Temperature

u: Velocity component tangent to the shock in the flow plane

v: Velocity component normal to the shock in the downstream direction

v_j: Velocity component in the Cartesian, $\hat{1}_j$, system

V: Velocity magnitude

w: $m\,\sin^2\beta$

x_i: Cartesian, $\hat{1}_i$, coordinate

x_i^*: Arbitrary point on a shock

X: $1 + \dfrac{\gamma-1}{2}w$

y: Radial or transverse coordinate; variable

Y: $\gamma w - \dfrac{\gamma-1}{2}$

Z: $w - 1$

GREEK

$\boldsymbol{\beta}$: Shock wave angle in the flow plane, relative to $\vec{V}_1$

$\boldsymbol{\beta'}, \boldsymbol{\beta_s}$: $d\beta/ds$

$\boldsymbol{\delta_i'}$: Angles that $\vec{V}_i$ have relative to the x-coordinate

$\boldsymbol{\Delta}$: Shock stand-off distance; Equation (8.27) determinant

$\boldsymbol{\gamma}$: Ratio of specific heats

$\boldsymbol{\varepsilon}$: Small positive constant

$\boldsymbol{\theta}$: Acute angle between $\vec{V}_1$ and $\vec{V}_2$

$\boldsymbol{\zeta_o}$: Streamline characteristic

$\boldsymbol{\zeta_-}$: Right-running characteristic

$\boldsymbol{\zeta_+}$: Left-running characteristic

$\boldsymbol{\kappa}$: Curvature

$\boldsymbol{\mu}$: Mach angle

$\boldsymbol{\xi_i}$: Orthogonal curvilinear coordinates

$\boldsymbol{\rho}$: Density

$\boldsymbol{\sigma}$: 0 for two-dimensional flow, 1 for axisymmetric flow

$\boldsymbol{\Sigma}$: Summation symbol

$\boldsymbol{\chi}$:
$$\frac{1}{V_1 \mid \nabla F \mid \cos\beta} = \frac{1}{\left(\sum K_j^2\right)^{1/2}}$$

$\boldsymbol{\psi}$: $F_{x2}^2 + F_{x3}^2$

$\boldsymbol{\omega}$: Vorticity

SUBSCRIPTS AND SUPERSCRIPTS

$()_a$: Flow plane

$()_b$: Plane normal to the shock and normal to the flow plane; body

$()_{cp}$: Crocco point

$()_n$: Normal to the shock

$()_s$: Derivative along the shock in the flow plane

$()_t$: Tip; transverse

$()_{tp}$: Thomas point

$()_1$: State just upstream of the shock

$()_2$: State just downstream of the shock

$()_o$: Stagnation value

$()^*$: Arbitrary point on the shock; sonic state just downstream of the shock; vector relative to a moving shock

$()'$: Angle measured relative to the x-coordinate

SPECIAL SYMBOLS

$\hat{1}_i$: Cartesian basis

$(\hat{})$: Unit vector

$(\bar{\ })$: Vector

$\mathbf{V}$: Del operator

$D()/Dt$: Substantial derivative

$(\tilde{\ })$: Dimensional parameter; angle measured counterclockwise relative to $\vec{V}_1$

$(\hat{\ })$: Denotes intrinsic coordinate parameter

Appendix B: Oblique Shock Wave Angle

Let β be the shock wave angle and θ be the velocity turn angle. Both are measured with respect to the velocity upstream of the shock, as pictured in Figure 2.1. The two angles are related by Equation (2.28), which is an explicit relation for θ. Because θ represents the usually known wall turn angle, an explicit, computer-friendly equation for β is desirable. The derivation of this relation is placed in an appendix because of its frequent usage in gas dynamics.

Thompson (1950) may have been the first to observe that Equation (2.28) could be written as a cubic in $\sin^2\beta$

$$\sin^6\beta - \left(\frac{M^2+2}{M^2} + \gamma \sin^2\theta\right)\sin^4\beta + \left\{\frac{2M^2+1}{M^4} + \left[\left(\frac{\gamma+1}{2}\right)^2 + \frac{\gamma-1}{M^2}\right]\sin^2\theta\right\}\sin^2\beta$$

$$-\frac{\cos^2\theta}{M^4} = 0 \tag{B.1}$$

It is analytically convenient, however, to recast this relation as a cubic in $\tan\beta$, with the result

$$F(\beta) = \left(1 + \frac{\gamma-1}{2}M^2\right)\tan\theta\ \tan^3\beta - \left(M^2-1\right)\tan^2\beta$$

$$+ \left(1 + \frac{\gamma+1}{2}M^2\right)\tan\theta\ \tan\beta + 1 = 0 \tag{B.2}$$

Observe that the coefficients in this polynomial are appreciably simpler than those in Equation (B.1). We assume $M > 1$ and $\theta > 0$, and note that

$$F(-\pi/2) = -\infty, \quad F(0) = 1, \quad F(\pi/2) = \infty,$$

From these values, we deduce that $F(\beta)$ has three real, unequal roots for an attached shock wave. The negative β root is not physical, while the two roots between $\beta = 0$ and $\pi/2$ correspond to the weak and strong shock solutions.

Equation (B.2) is next recast into the standard form

$$x^3 + ax + b = 0 \tag{B.3}$$

for solving a cubic equation, where

$$x = \tan\beta - \frac{M^2 - 1}{3\left(1 + \frac{\gamma - 1}{2}M^2\right)\tan\theta}$$

(B.4)

$$a = \frac{3\left(1 + \frac{\gamma - 1}{2}M^2\right)\left(1 + \frac{\gamma + 1}{2}M^2\right)\tan^2\theta - \left(M^2 - 1\right)^2}{3\left(1 + \frac{\gamma - 1}{2}M^2\right)^2 \tan^2\theta}$$

$$b = \frac{-2\left(M^2 - 1\right)^3 + 18\left(1 + \frac{\gamma - 1}{2}M^2\right)\left(1 + \frac{\gamma - 1}{2}M^2 + \frac{\gamma + 1}{4}M^4\right)\tan^2\theta}{27\left(1 + \frac{\gamma - 1}{2}M^2\right)^3 \tan^3\theta}$$

Because the roots are real and unequal, the trigonometric solution of a cubic equation is particularly convenient. This solution requires the quantity:

$$\chi = \left(-\frac{27b^2}{4a^3}\right)^{1/2} = \frac{\left(M^2 - 1\right)^3 - 9\left(1 + \frac{\gamma - 1}{2}M^2\right)\left(1 + \frac{\gamma - 1}{2}M^2 + \frac{\gamma + 1}{4}M^4\right)\tan^2\theta}{\left[\left(M^2 - 1\right)^2 - 3\left(1 + \frac{\gamma - 1}{2}M^2\right)\left(1 + \frac{\gamma + 1}{2}M^2\right)\tan^2\theta\right]^{3/2}}$$

(B.5)

The three solutions of Equation (B.3) are contained among the six expressions

$$\pm\left(-\frac{4a}{3}\right)^{1/2}\cos\frac{\phi}{3}, \quad \pm\left(-\frac{4a}{3}\right)^{1/2}\cos\left(\frac{\phi + 2\pi}{3}\right), \quad \pm\left(-\frac{4a}{3}\right)^{1/2}\cos\left(\frac{\phi + 4\pi}{3}\right)$$

where $\phi = \cos^{-1}\chi$.

In particular, the weak and strong solutions are given by

$$x_{weak} = \left(-\frac{4a}{3}\right)^{1/2}\cos\frac{\phi + 4\pi}{3}, \quad x_{strong} = \left(-\frac{4a}{3}\right)^{1/2}\cos\frac{\phi}{3}$$

where

$$\left(-\frac{4a}{3}\right)^{1/2} = \frac{2\left[\left(M^2 - 1\right)^2 - 3\left(1 + \frac{\gamma - 1}{2}M^2\right)\left(1 + \frac{\gamma + 1}{2}M^2\right)\tan^2\theta\right]^{1/2}}{3\left(1 + \frac{\gamma - 1}{2}M^2\right)\tan\theta}$$

With the aid of Equation (B.5), a computationally convenient form for β is

$$\lambda = \left[\left(M^2 - 1\right)^2 - 3\left(1 + \frac{\gamma - 1}{2}M^2\right)\left(1 + \frac{\gamma + 1}{2}M^2\right)\tan^2\theta\right]^{1/2} \qquad \text{(B.6a)}$$

$$\chi = \frac{\left(M^2 - 1\right)^3 - 9\left(1 + \frac{\gamma - 1}{2}M^2\right)\left(1 + \frac{\gamma - 1}{2}M^2 + \frac{\gamma + 1}{4}M^4\right)\tan^2\theta}{\lambda^3} \qquad \text{(B.6b)}$$

$$\tan\beta = \frac{M^2 - 1 + 2\lambda\cos\left[\left(4\pi\delta + \cos^{-1}\chi\right)/3\right]}{3\left(1 + \frac{\gamma - 1}{2}M^2\right)\tan\theta} \qquad \text{(B.6c)}$$

where the angle $(4\pi\delta + \cos^{-1}\chi)/3$ is in radians. The strong shock solution is provided by $\delta = 0$, while $\delta = 1$ yields the weak shock solution, and $|\chi| \leq 1$ for an attached shock.

The author discovered the solution to Equation (B.6) in March 1991. It was quickly submitted and accepted for publication by the AIAA (American Institute of Aeronautics and Astronautics) journal. Before it was scheduled to appear, however, it was learned that the solution, in a different form, had already been published. In fact, it has repeatedly appeared in journals (e.g., see Mascitti 1969; Wolf 1993).

REFERENCES

Mascitti, V. R., A Closed-Form Solution to Oblique Shock-Wave Properties, *J. Aircraft* 6, 66 (1969).

Thompson, M. J., A Note on the Calculation of Oblique Shock-Wave Characteristics, *J. Aeron. Sci.* 17, 774 (1950).

Wolf, T., Comment on Approximate Formula of Weak Oblique Shock Wave Angle, *AIAA J.* 31, 1363 (1993).

Appendix C: Method-of-Characteristics for a Single, First-Order Partial Differential Equation

There are many ways to introduce the method-of-characteristics (MOC). Here interest is limited to a single, first-order, linear or quasilinear partial differential equation (PDE). Our approach is thus specifically tailored for the task at hand.

For purpose of generality, we consider an inhomogeneous equation for the dependent variable f

$$\sum_{i=0}^{n-1} a_i \frac{\partial f}{\partial x_i} + a_n = 0 \tag{C.1}$$

where n is a positive integer. This equation is assumed to be quasilinear, in which case $a_0, \ldots, a_n$ can depend on the x_j and f, but not on any derivative of f. We further simplify the equation by noting that if f is a solution, then

$$G(x_0, \ldots, x_n) = f(x_0, \ldots, x_{n-1}) + x_n \tag{C.2}$$

is a solution of the homogeneous equation

$$\sum_{i=0}^{n} a_i \frac{\partial G}{\partial x_i} = 0 \tag{C.3}$$

Thus, by adding a new independent variable, x_n, the inhomogeneous term in Equation (C.1) is incorporated into Equation (C.3).

C.1 GENERAL SOLUTION

Observe that $G = $ constant is a solution of Equation (C.3). This constant may be taken as zero. We therefore seek a solution with the form

$$G(x_0, \ldots, x_n) = 0 \tag{C.4}$$

The remainder of the section provides this solution.

It is conceptually convenient to introduce an $(n + 1)$-dimensional Cartesian space that has an orthonormal basis $\hat{1}_i$. Thus, the gradient of G is

$$\nabla G = \sum_{i=0}^{n} \frac{\partial G}{\partial x_1} \hat{1}_i \tag{C.5}$$

can be defined that is based on the a_i coefficients

$$\vec{A} = \sum_{i=0}^{n} a_i \hat{1}_i$$

Hence, Equation (C.3) becomes

$$\vec{A} \cdot \nabla G = 0 \tag{C.6}$$

Equation (C.4) represents a surface in an $(n + 1)$-dimensional space, and the gradient ∇G is everywhere normal to this surface. On the other hand, $\vec{A}$ is perpendicular to ∇G and therefore $\vec{A}$ is tangent to the surface. Thus, the solution of Equation (C.3) or (C.6) is a surface that is tangent to $\vec{A}$.

Consider a characteristic curve that lies on the surface given by Equation (C.4) and everywhere is tangent to $\vec{A}$. The surface can be viewed as consisting of an infinite number of these curves. Moreover, each of these curves constitutes a solution of Equation (C.3).

We need to construct a curve in the $(n + 1)$-dimensional space whose coordinates are $x_0, \ldots, x_n$. For example, in three dimensions a curve is determined by the intersection of two surfaces. More generally, the characteristic curve we seek is determined by the intersection of the n surfaces:

$$u^{(0)}(x_0, \ldots, x_n) = c_0$$

$$u^{(1)}(x_0, \ldots, x_n) = c_1$$

$$\vdots \tag{C.7}$$

$$u^{(n-1)}(x_0, \ldots, x_n) = c_{n-1}$$

where the c_j are constants, and the first equation is sometimes written as $u = c$. We have a different curve for each choice of the c_j.

Because $\vec{A}$ is tangent to a characteristic curve, the differential change in the x_i along such a curve must stand in the same relationship to each other as the corresponding component of $\vec{A}$. Thus, on a characteristic curve, we have

$$\frac{dx_0}{a_0} = \frac{dx_1}{a_1} = \ldots = \frac{dx_n}{a_n} \tag{C.8}$$

As noted, G is constant along a characteristic curve. We therefore see from Equation (C.2) that dx_n can be replaced with $-df$. This change is usually convenient, because the a_i are functions of $x_0, \ldots, x_{n-1}$ and f. The equations in Equation (C.8) are n-coupled, first-order ordinary differential equations (ODEs) that relate the x_i along a characteristic curve. The unique solution of these equations is provided by Equation (C.7), where the c_j are constants of integration. We thus have reduced the problem of solving a first-order PDE to solving n-coupled ODEs. As will become apparent, this reduction is advantageous whether Equation (C.3) is to be solved analytically or numerically.

We now see why Equation (C.3) is limited to being quasilinear. If one of the a_i depended on a derivative of f, then one of Equation (C.8) would not be an ODE, and the theory would collapse. Normally, the MOC applies only to hyperbolic equations. For Equation (C.3), this qualification is unnecessary. The only essential restriction is that it be quasilinear.

Note that $\vec{A} \cdot \nabla G$ is also the derivative of G along a characteristic curve. Equation (C.6) therefore means that G has a constant value along any particular characteristic curve. For this to be so, G can depend on the x_i only in combinations such as $dG = 0$ along any characteristic curve. However, the $u^{(j)}$ depend on the x_i but have a constant value along any characteristic curve. Consequently, G is an arbitrary function of the $u^{(j)}$. The general solution of Equation (C.3) is thus

$$G(u^{(0)}, u^{(1)}, \ldots, u^{(n-1)}) = 0 \tag{C.9}$$

If one or more of the a_i depend on f, or if $a_n \neq 0$, then f explicitly appears in the $u^{(j)}$, and Equation (C.9) is a solution of Equation (C.1). On the other hand, if none of the a_i involves f and $a_n = 0$, then the general solution of Equation (C.1) can be written as

$$f = f(u^{(0)}, u^{(1)}, \ldots, u^{(n-2)}) \tag{C.10}$$

C.2   DISCUSSION

We verify that Equation (C.9) is a solution of Equation (C.3) by first evaluating $du^{(j)}$ with the aid of Equation (C.8):

$$du^{(j)} = \sum_{i=0}^{n} \frac{\partial u^{(j)}}{\partial x_i} dx_i = \frac{dx_0}{a_0} \sum_{i=0}^{n} a_i \frac{\partial u^{(j)}}{\partial x_i}$$

where we assume one a_i, say a_0, is nonzero. We next obtain

$$dG = \sum_{j=0}^{n-1} \frac{\partial G}{\partial u^{(j)}} du^{(j)} = \frac{dx_0}{a_0} \sum_{j=0}^{n-1} \frac{\partial G}{\partial u^{(j)}} \sum_{j=0}^{n} a_i \frac{\partial u^{(j)}}{\partial x_i}$$

$$= \frac{dx_0}{a_0} \sum_{j=0}^{n} a_i \sum_{i=0}^{n-1} \frac{\partial G}{\partial u^{(j)}} \frac{\partial u^{(j)}}{\partial x_i} = \frac{dx_0}{a_0} \sum_{i=0}^{n} a_i \frac{\partial G}{\partial x_i} = 0$$

in accordance with Equation (C.3).

The functional form of G is determined by an initial, or boundary, condition. Without loss of generality, this condition may be specified at $x_0 = 0$ as

$$G_0 = G[u^{(0)}(0, x_1, \ldots, x_n), \ldots, u^{(n-1)}(0, x_1, \ldots, x_n)]$$

where G_0 is the prescribed relation for G at $x_0 = 0$.

As we have mentioned, the unique solution of Equation (C.8) can be written as Equation (C.7). An analytical solution of Equation (C.8) may require inverting some of Equation (C.7). For example, suppose $n = 3$, and we have obtained a solution, $u = c$, to

$$\frac{dx_0}{a_0} = \frac{dx_1}{a_1}$$

Further, suppose a_2 depends on x_0, x_1, and x_2. If $u = c$ can be explicitly solved for x_0, we would then integrate

$$\frac{dx_1}{a_1} = \frac{dx_2}{a_2}$$

with x_0 eliminated. Similarly, if $u = c$ is more readily solved for x_1, we could obtain $u^{(1)}$ by integrating

$$\frac{dx_0}{a_0} = \frac{dx_2}{a_2}$$

instead. In either case, the elimination of x_0 (or x_1) from the dx_2 equation is consistent with obtaining a simultaneous solution of Equation (C.8).

C.3 ILLUSTRATIVE EXAMPLE

As an example, the general solution to

$$xz\frac{\partial z}{\partial x} + yz\frac{\partial z}{\partial y} = xy$$

is found. We first solve the characteristic equations:

$$\frac{dx}{xz} = \frac{dy}{yz} = \frac{dz}{xy}$$

From the leftmost equation, we have

$$\frac{dx}{x} = \frac{dy}{y}$$

which integrates to

$$u = \frac{y}{x} = c$$

For a second equation, we use

$$\frac{dx}{z} = \frac{dz}{y}$$

or by the elimination of y

$$cxdx = zdz$$

$$cx^2 = z^2 - c_1$$

$$\left(\frac{y}{x}\right) x^2 = z^2 - c_1$$

$$u^{(1)} = z^2 - xy = c_1$$

Hence, the general solution to the PDE is

$$g(z^2 - xy, y/x) = 0$$

which is readily verified by direct substitution. An alternate form for the solution can be written as

$$z^2 - xy = g(y/x)$$

or as

$$z = \pm[xy + g(y/x)]^{1/2}$$

where g is an arbitrary function of its argument.

Appendix D: Orthogonal Basis Derivatives

D.1 UNITARY BASIS DERIVATIVES

Γ_{ij}^k = Christoffel symbol

$$= \Gamma_{ji}^k = \frac{\partial \vec{e}_i}{\partial q^j} \cdot \vec{e}^k = \frac{g^{kr}}{2}\left(\frac{\partial g_{jr}}{\partial q^i} + \frac{\partial g_{ir}}{\partial q^j} - \frac{\partial g_{ij}}{\partial q^r}\right)$$

$$\frac{\partial \vec{e}_i}{\partial q^j} = \Gamma_{ij}^k \vec{e}^k, \qquad \frac{\partial \vec{e}^k}{\partial q^j} = -\Gamma_{ij}^k \vec{e}^i$$

$$d\vec{e}_i = \Gamma_{ij}^k dq^j \vec{e}_k, \qquad d\vec{e}^k = -\Gamma_{ij}^k dq^j \vec{e}^i$$

D.2 ORTHOGONAL FORMS FOR THE CHRISTOFFEL SYMBOL

Case	i,j,k	Γ_{ij}^k
1	All different	0
2	$k=j,\quad i \ne j$	$\dfrac{1}{h_j}\dfrac{\partial h_j}{\partial q^i}$ (no sum)
3	$k=1 \quad i \ne j$	$\dfrac{1}{h_i}\dfrac{\partial h_i}{\partial q^j}$ (no sum)
4	$k \ne j \quad i = j$	$\dfrac{h_i}{h_k^2}\dfrac{\partial h_i}{\partial q^k}$ (no sum)
5	$i=j=k$	$\dfrac{1}{h_i}\dfrac{\partial h_i}{\partial q^i}$ (no sum)

$$\Gamma_{ij}^k = \frac{1}{h_k^2}\left(\delta_{jk}h_j\frac{\partial h_j}{\partial q^i} + \delta_{ik}h_i\frac{\partial h_i}{\partial q^j} - \delta_{ij}h_i\frac{\partial h_i}{\partial q^k}\right) \quad \text{(no sum)}$$

$$\frac{\partial \hat{e}_i}{\partial q^j} = \sum_{k \ne i}\left(\frac{\delta_{jk}}{h_i}\frac{\partial h_j}{\partial q^i} - \frac{\delta_{ij}}{h_k}\frac{\partial h_i}{\partial q^k}\right)\hat{e}_k$$

$$d\hat{e}_i = \sum_{j \ne i}\left(\frac{1}{h_i}\frac{\partial h_j}{\partial q^i}dq^j - \frac{1}{h_j}\frac{\partial h_i}{\partial q^j}dq^i\right)\hat{e}_j$$

Appendix E: Conditions on the Downstream Side of a Two-Dimensional or Axisymmetric Shock with a Uniform Freestream

E.1 JUMP CONDITIONS

$$m \equiv M_1^2, \quad w \equiv m\sin^2\beta, \quad X \equiv 1 + \frac{\gamma-1}{2}w, \quad Y \equiv \gamma w - \frac{\gamma-1}{2}, \quad Z \equiv w - 1$$

$$A \equiv \frac{\gamma+1}{2}\frac{m\ \sin\beta\cos\beta}{X}, \quad B \equiv 1 + A^2$$

$$\frac{u_2}{V_1} = \cos\beta$$

$$\frac{v_2}{V_1} = \frac{2}{\gamma+1}\frac{X}{m\sin\beta}$$

$$\frac{V_2}{V_1} = \frac{1}{V_1}\left(u_2^2 + v_2^2\right)^{1/2} = \frac{2}{\gamma+1}\frac{XB^{1/2}}{(mw)^{1/2}}$$

$$\frac{p_2}{p_1} = \frac{2}{\gamma+1}Y$$

$$\frac{\rho_2}{\rho_1} = \frac{\gamma+1}{2}\frac{w}{X}$$

$$\frac{T_2}{T_1} = \left(\frac{2}{\gamma+1}\right)^2\frac{XY}{w}$$

$$M_2^2 = \frac{X}{Y}B$$

$$1+\frac{\gamma-1}{2}M_2^2=\left(\frac{\gamma+1}{2}\right)^2\frac{\left[1+(\gamma-1)m/2\right]w}{XY}$$

$$\frac{p_{O,2}}{p_1}=\frac{2}{(\gamma+1)}\left(1+\frac{\gamma-1}{2}M_2^2\right)^{\gamma/(\gamma-1)}Y$$

$$\tan\theta=\frac{1}{\tan\beta}\frac{Z}{(\gamma+1)m/2-Z}$$

$$\sin\theta=\frac{Z\cos\beta}{XB^{1/2}}$$

$$\sin(\beta-\theta)=\frac{1}{B^{1/2}}$$

$$\cos(\beta-\theta)=\frac{A}{B^{1/2}}$$

E.2   TANGENTIAL DERIVATIVES

$$\frac{1}{V_1}\left(\frac{\partial u}{\partial s}\right)_2=-\beta'\sin\beta$$

$$\frac{1}{V_1}\left(\frac{\partial v}{\partial s}\right)_2=-\frac{2}{\gamma+1}\left(1-\frac{\gamma-1}{2}w\right)\frac{\beta'\cos\beta}{w}$$

$$\frac{1}{p_1}\left(\frac{\partial p}{\partial s}\right)_2=\frac{4\gamma}{\gamma+1}\beta'm\sin\beta\cos\beta$$

$$\frac{1}{\rho_1}\left(\frac{\partial\rho}{\partial s}\right)_2=2\frac{A}{X}\beta'$$

$$\frac{1}{T_1}\left(\frac{\partial T}{\partial s}\right)_2=\frac{4(\gamma-1)}{(\gamma+1)^2}\frac{\left(1+\gamma w^2\right)}{w}\frac{\beta'}{\tan\beta}$$

$$\left(\frac{\partial M^2}{\partial s}\right)_2 = -(\gamma+1)\left(1+\frac{\gamma-1}{2}m\right)(1+\gamma w^2)\frac{A}{XY^2}\beta'$$

$$\frac{1}{p_1}\left(\frac{\partial p_o}{\partial s}\right)_2 = -\frac{2\gamma}{\gamma+1}\left(1+\frac{\gamma-1}{2}M_2^2\right)^{\gamma/(\gamma-1)}\frac{\beta' Z^2}{X\tan\beta}$$

$$\left(\frac{\partial\theta}{\partial s}\right)_2 = \frac{\dfrac{\gamma+1}{2}m(1+w)+1-2w-\gamma w^2}{X^2 B}\beta'$$

$$\left(\frac{\partial\mu}{\partial s}\right)_2 = -\frac{1}{2M_2^2\left(M_2^2-1\right)^{1/2}}\left(\frac{\partial M^2}{\partial s}\right)_2$$

E.3   NORMAL DERIVATIVES

$$\frac{1}{V_1}\left(\frac{\partial u}{\partial n}\right)_2 = \frac{g_1\beta'\cos\beta}{X}$$

$$\frac{1}{V_1}\left(\frac{\partial v}{\partial n}\right)_2 = \frac{1}{m\sin\beta}\left[g_2-\frac{2}{\gamma+1}m(1+3w)\right]\frac{\beta'}{Z}-\left(\frac{2}{\gamma+1}\right)^2\frac{Y}{m\sin\beta}\frac{\sigma\cos\beta}{y}$$

$$\frac{1}{p_1}\left(\frac{\partial p}{\partial n}\right)_2 = \frac{\gamma}{\gamma+1}\frac{(mg_5+g_6)\beta'}{XZ}+\left(\frac{2}{\gamma+1}\right)^2\gamma Y\frac{\sigma\cos\beta}{y}$$

$$\frac{1}{\rho_1}\left(\frac{\partial\rho}{\partial n}\right)_2 = \frac{w}{X}\left[(mg_3+g_4)\frac{\beta'}{X^2 Z}+\frac{\sigma\cos\beta}{y}\right]$$

$$\frac{1}{T_1}\left(\frac{\partial T}{\partial n}\right)_2 = \left(\frac{2}{\gamma+1}\right)^2\frac{X}{w}\left\{\left[\gamma X(mg_5+g_6)-\frac{4}{\gamma+1}Y(mg_3+g_4)\right]\frac{\beta'}{2X^2 Z}\right.$$
$$\left.+2\left(\frac{\gamma-1}{\gamma+1}\right)Y\frac{\sigma\cos\beta}{y}\right\}$$

$$\left(\frac{\partial M^2}{\partial n}\right)_2 = -(\gamma+1)\frac{\left(1+\dfrac{\gamma-1}{2}m\right)w}{XY^2}\left[\frac{(mg_7+g_8)\beta'}{4X^2Z}+Y\frac{\sigma\cos\beta}{y}\right]$$

$$\frac{1}{p_1}\left(\frac{\partial p_o}{\partial n}\right)_2 = \gamma m\left(1+\frac{\gamma-1}{2}M_2^2\right)^{\gamma/(\gamma-1)}\left(\frac{Z\cos\beta}{X}\right)^2\beta'$$

E.4 $g_i(\gamma,w)$

$$g_1 = \frac{1}{\gamma+1}\left[-(\gamma+5)+(3-\gamma)w\right]$$

$$g_2 = \frac{2}{(\gamma+1)^2}\left[(\gamma-1)-2(\gamma-1)w+(5\gamma+3)w^2\right]$$

$$g_3 = \frac{\gamma+1}{4}\left[2(\gamma+2)+(3-\gamma)w+3(\gamma-1)w^2\right]$$

$$g_4 = \frac{1}{2}\left[2-(\gamma^2+\gamma+6)w+(\gamma^2-4\gamma+1)w^2-(\gamma-1)(2\gamma+1)w^3\right]$$

$$g_5 = -(\gamma-1)+(\gamma+5)w+2(2\gamma-1)w^2$$

$$g_6 = \frac{2}{\gamma+1}\left[-(\gamma-1)+2(2\gamma-1)w+(\gamma^2-7\gamma-2)w^2-(3\gamma^2-1)w^3\right]$$

$$g_7 = \frac{\gamma+1}{2}\left[4-(\gamma-1)(\gamma+3)w+(\gamma^2+18\gamma-3)w^2-4\gamma(2-\gamma)w^3\right]$$

$$g_8 = -2(\gamma-1)+2(\gamma-1)(3-\gamma)w+(9\gamma^2-14\gamma+1)w^2$$
$$+(\gamma^3-17\gamma^2-\gamma+1)w^3+\gamma(-3\gamma^2+4\gamma+3)w^4$$

Appendix F: Conditions on the Downstream Side of a Two-Dimensional or Axisymmetric Shock When the Upstream Flow Is Nonuniform

F.1 JUMP CONDITIONS

$$m \equiv M_1^2, \quad w \equiv m\,\sin^2\beta, \quad X \equiv 1 + \frac{\gamma-1}{2}w, \quad Y \equiv \gamma w - \frac{\gamma-1}{2}, \quad Z \equiv w - 1,$$

$$A \equiv \frac{\gamma+1}{2}\,\frac{m\sin\beta\,\cos\beta}{X}, \quad B \equiv 1 + A^2$$

$$q_j = (u,\,v,\,p,\,\rho)_2 = f_j$$

$$f_1 = \cos\beta, \quad f_2 = \frac{2}{\gamma+1}\,\frac{X}{m\sin\beta}, \quad f_3 = \frac{2}{\gamma+1}\,Y, \quad f_4 = \frac{\gamma+1}{2}\,\frac{w}{X}$$

F.2 TANGENTIAL DERIVATIVES

$$q_{js} = (u_{2s},\,v_{2s},\,p_{2s},\,\rho_{2s})$$

$$\chi_i = (M_{1s}\,/\,M_1,\,V_{1s},\,p_{1s},\,\rho_{1s},\,\beta_s)$$

$$q_{js} = \sum_{i=1}^{5} g_{ji}\chi_i$$

$$q_1 = u_2:$$

$$g_{11} = 0, \quad g_{12} = \cos\beta, \quad g_{13} = 0, \quad g_{14} = 0, \quad g_{15} = -\sin\beta$$

$$q_2 = v_2:$$

$$g_{21} = -\frac{4}{\gamma+1}\,\frac{1}{m\sin\beta}, \quad g_{22} = \frac{2}{\gamma+1}\,\frac{X}{m\sin\beta}, \quad g_{23} = 0, \quad g_{24} = 0,$$

$$g_{25} = -\frac{2}{\gamma+1}\frac{1-\dfrac{\gamma-1}{2}\,w}{w}\cos\beta$$

$$q_3 = p_2:$$

$$g_{31} = \frac{4\gamma}{\gamma+1}\,w, \quad g_{32} = 0, \quad g_{33} = \frac{2}{\gamma+1}\,Y, \quad g_{34} = 0, \quad g_{35} = \frac{8\gamma}{(\gamma+1)^2}\,XA$$

$$q_4 = \rho_2:$$

$$g_{41} = (\gamma+1)\frac{w}{X^2}, \quad g_{42} = 0, \quad g_{43} = 0, \quad g_{44} = \frac{\gamma+1}{2}\frac{w}{X}, \quad g_{45} = \frac{2A}{X}$$

F.3 NORMAL DERIVATIVES

$$q_{jn} = \left(u_{2n},\, v_{2n},\, p_{2n},\, \rho_{2n}\right)$$

$$\chi_i' = \left(u_{2s},\, v_{2s},\, p_{2s},\, \rho_{2s},\, \beta_s',\, \frac{\sigma}{y}\,\alpha_1\right)$$

$$q_{jn} = \sum_{i=1}^{6} h_{ji}\chi_i'$$

$$\Delta = \frac{p_2}{m\rho_2} - v_2^2 = \frac{2}{\gamma+1}\frac{XZ}{mw}$$

$$\alpha_1 = \frac{2}{\gamma+1}\frac{X}{m\sin\beta}\left(A\sin\beta' - \cos\beta'\right)$$

$$q_1 = u_2:$$

$$h_{11} = -\frac{u_2}{v_2}, \quad h_{12} = 0, \quad h_{13} = -\frac{1}{\gamma m\rho_2 v_2}, \quad h_{14} = 0, \quad h_{15} = -u_2, \quad h_{16} = 0$$

$$q_2 = v_2:$$

$$h_{21} = \frac{1}{\Delta}\left[-\frac{p_2}{m\rho_2} + (\gamma-1)u_2^2\right], \quad h_{22} = \frac{\gamma u_2 v_2}{\Delta}, \quad h_{23} = \frac{\gamma-1}{\gamma}\frac{u_2}{\Delta m\rho_2},$$

$$h_{24} = -\frac{p_2 u_2}{\Delta m\rho_2^2}, \quad h_{25} = -\frac{v_2}{\Delta}\left(\frac{p_2}{m\rho_2} + u_2^2\right), \quad h_{26} = -\frac{p_2}{\Delta m\rho_2}$$

$$q_3 = p:$$

$$h_{31} = \frac{\gamma m \rho_2 v_2}{\Delta} \left[\frac{p_2}{m \rho_2} - (\gamma - 1) u_2^2 \right], \quad h_{32} = -\frac{\gamma m \rho_2 \, u_2}{\Delta} \left[\frac{p_2}{m \rho_2} + (\gamma - 1) v_2^2 \right],$$

$$h_{33} = -\frac{(\gamma - 1) u_2 v_2}{\Delta}, \quad h_{34} = \frac{\gamma p_2 u_2 v_2}{\Delta \rho_2}, \quad h_{35} = \frac{\gamma p_2 \left(u_2^2 + v_2^2 \right)}{\Delta}, \quad h_{36} = \frac{\gamma p_2 v_2}{\Delta}$$

$$q_4 = \rho:$$

$$h_{41} = \frac{p_2}{\Delta v_2} \left[-(\gamma - 1) u_2^2 + v_2^2 \right], \quad h_{42} = -\frac{\gamma p_2 \, u_2}{\Delta}, \quad h_{43} = -\frac{\gamma - 1}{\gamma} \frac{u_2}{\Delta m v_2},$$

$$h_{44} = \frac{u_2 v_2}{\Delta}, \quad h_{45} = \frac{p_2 \left(u_2^2 + v_2^2 \right)}{\Delta}, \quad h_{46} = \frac{p_2 v_2}{\Delta}$$

Appendix G: Operator Formulation

Let $\hat{\mathfrak{l}}_j$ and $\hat{e}_i$ be two right-handed, Cartesian bases related by

$$\hat{e}_i = \sum_{j=1}^{3} \alpha_{ij}\, \hat{\mathfrak{l}}_j, \quad i = 1, 2, 3 \tag{G.1}$$

where

$$\sum_j \alpha_{ij}^2 = 1, \quad i = 1, 2, 3 \tag{G.2}$$

and the α_{ij} $(=\hat{e}_i \cdot \hat{\mathfrak{l}}_j)$ are direction cosines. Write the position vector as

$$\vec{r} = \sum x_j \hat{\mathfrak{l}}_j = \sum y_i \hat{e}_i \tag{G.3}$$

where the x_j and y_i are the $\vec{r}$ coordinates in the two systems. We now have

$$\frac{\partial \vec{r}}{\partial y_i} = \sum \frac{\partial x_j}{\partial y_i}\, \hat{\mathfrak{l}}_j = \hat{e}_i = \sum \alpha_{ij}\, \hat{\mathfrak{l}}_j \tag{G.4}$$

and, consequently,

$$\frac{\partial x_j}{\partial y_i} = \alpha_{ij} \tag{G.5}$$

The differential of $\vec{r}$ with respect to x_j yields

$$\frac{\partial \vec{r}}{\partial x_j} = \sum \frac{\partial y_i}{\partial x_j}\, \hat{e}_i = \hat{\mathfrak{l}}_j \tag{G.6}$$

The inversion of Equation (G.1) is written as

$$\hat{\mathfrak{l}}_j = \sum \beta_{ji}\hat{e}_i, \quad \sum_i \beta_{ji}^2 = 1 \tag{G.7}$$

where

$$\beta_{ji} = \frac{\partial y_i}{\partial x_j} \tag{G.8}$$

The α and β matrices are orthogonal (Goldstein 1950)—that is,

$$\beta = \alpha^{-1} = \alpha^t \tag{G.9}$$

where α^t is the transpose of α. Hence, we have

$$\left(\frac{\partial y_i}{\partial x_j}\right)_{x_k} = \left(\frac{\partial x_j}{\partial y_i}\right)_{y_k} = \alpha_{ij} = \beta_{ji} \tag{G.10}$$

where the subscript on the derivatives indicates the fixed variable.

When applied to Equations (9.10), (9.2), and (9.7b), we obtain

$$\hat{e}_1 = \hat{t} = \frac{\chi}{|\nabla F|}\sum L_j \,\hat{1}_j, \quad \hat{e}_2 = \hat{n} = \frac{1}{|\nabla F|}\sum F_{x_j} \,\hat{1}_j, \quad \hat{e}_3 = \hat{b} = -\chi\sum K_j \,\hat{1}_j \tag{G.11}$$

and

$$\alpha_{1j} = \frac{\chi}{|\nabla F|}L_j, \quad \alpha_{2j} = \frac{F_{x_j}}{|\nabla F|}, \quad \alpha_{3j} = -\chi K_j \quad j = 1, 2, 3 \tag{G.12}$$

With

$$y_1 = s, \quad y_2 = n, \quad y_3 = b \tag{G.13}$$

Equations (G.10) and (G.12) provide

$$\frac{\partial s}{\partial x_i} = \frac{\chi}{|\nabla F|}L_i, \quad \frac{\partial n}{\partial x_i} = \frac{F_{x_i}}{|\nabla F|}, \quad \frac{\partial b}{\partial x_i} = -\chi K_i \tag{G.14}$$

The transformation of x_i partial derivatives is given by the chain rule

$$\frac{\partial}{\partial x_i} = \frac{\chi}{|\nabla F|}L_i\frac{\partial}{\partial s} + \frac{F_{x_i}}{|\nabla F|}\frac{\partial}{\partial n} - \chi K_i\frac{\partial}{\partial b} \tag{G.15}$$

This relation should not be confused with the S_a and S_b derivative operators discussed in Section 9.4.

For the inverse transformation, again use Equation (G.10)

$$\frac{\partial x_j}{\partial n} = \frac{F_{x_j}}{|\nabla F|}, \quad \frac{\partial x_j}{\partial s} = \frac{\chi}{|\nabla F|}L_j, \quad \frac{\partial x_j}{\partial b} = -\chi K_j \tag{G.16}$$

for the derivatives

$$\frac{\partial}{\partial s} = \sum \frac{\partial x_j}{\partial s}\frac{\partial}{\partial x_j} = \frac{\chi}{|\nabla F|}\sum L_j\frac{\partial}{\partial x_j} \tag{G.17a}$$

$$\frac{\partial}{\partial b} = \sum \frac{\partial x_j}{\partial b}\frac{\partial}{\partial x_j} = -\chi\sum K_j\frac{\partial}{\partial x_j} \tag{G.17b}$$

$$\frac{\partial}{\partial n} = \sum \frac{\partial x_j}{\partial n}\frac{\partial}{\partial x_j} = \frac{1}{|\nabla F|}\sum F_{x_j}\frac{\partial}{\partial x_j} \tag{G.17c}$$

Equation (G.17c) cannot be used for surface parameters, such as M_1 or β, because their derivatives with respect to n are zero. The consistency of Equations (G.15) and (G.17) can be verified with the use of Equation (9.13). For instance, we check $\partial()/\partial s$:

$$
\frac{\partial}{\partial s} = \frac{\chi}{|\nabla F|} \sum L_j \frac{\partial}{\partial x_j} = \frac{\chi}{|\nabla F|} \sum L_j \left(\frac{F_{x_j}}{|\nabla F|} \frac{\partial}{\partial n} + \frac{\chi}{|\nabla F|} L_j \frac{\partial}{\partial s} - \chi K_j \frac{\partial}{\partial b} \right)
$$

$$
= \frac{\chi}{|\nabla F|^2} \left(\sum F_{x_j} L_j \right) \frac{\partial}{\partial n} + \frac{\chi^2}{|\nabla F|^2} \left(\sum L_j^2 \right) \frac{\partial}{\partial s} - \frac{\chi^2}{|\nabla F|} \left(\sum K_j L_j \right) \frac{\partial}{\partial b} = \frac{\partial}{\partial s} \quad \text{(G.18)}
$$

A similar check holds for $\partial()/\partial b$ and $\partial()/\partial n$. Equations (G.17a,b) provide shock surface derivatives that apply on both sides of the shock. For instance, Equation (9.54a) for $(\partial w/\partial s)/w$ holds at state 1 even when the upstream flow is uniform, because there is a β derivative contribution. This also can be checked for the upstream side using the source flow model of Section 8.7 (Problem 23). Moreover, it is easy to show that

$$
\nabla = \sum \hat{\imath}_j \frac{\partial}{\partial x_j} = \hat{t} \frac{\partial}{\partial s} + \hat{n} \frac{\partial}{\partial n} + \hat{b} \frac{\partial}{\partial b} \quad \text{(G.19)}
$$

Equation (G.17c) does not circumvent the need of the Euler equations for the evaluation of normal derivatives, such as $(\partial p/\partial n)_2$. Equation (G.17c) simply replaces this derivative with the unknown $(\partial p/\partial x_j)_2$ derivatives. When the freestream is uniform, derivatives, such as $(\partial p/\partial s)_1$ and $(\partial M_1/\partial b)$, are clearly zero. In this case, Equation (G.17a,b) reduce to Equations (I.18) and (I.19) in Appendix I. For a surface parameter, such as β, Equation (G.17a,b) are useful, because β_{xi} is known, see Equations (9.35), (I.6), or (J.5). On the other hand, a jump parameter, say p, is given by (see Appendix E)

$$
\left(\frac{\partial p}{\partial s} \right)_2 = \frac{2}{\gamma+1} \left[\left(\frac{\partial p}{\partial s} \right)_1 Y + p_1 \, \gamma \, \frac{\partial w}{\partial s} \right]
$$

$$
\frac{1}{p_1} \left(\frac{\partial p}{\partial s} \right)_2 = \frac{2}{\gamma+1} \left(\frac{\chi}{|\nabla F|} Y \sum L_j \frac{p_{1x_j}}{p_1} + \gamma \frac{\partial w}{\partial s} \right) \quad \text{(G.20)}
$$

where $(\partial p/\partial s)_1$ is given by Equation (G.17a), and where, like M_{1x_j}, p_{1x_j} is presumed to be known. In a similar manner, we obtain

$$
\frac{1}{p_1} \left(\frac{\partial p}{\partial b} \right)_2 = - \frac{2}{\gamma+1} \left(\chi \, Y \sum K_j \frac{p_{1x_j}}{p_1} + \gamma \frac{\partial w}{\partial b} \right) \quad \text{(G.21a)}
$$

The s and b derivatives of w are provided by Equations (9.54) or (I.20) or (I.21) when the upstream flow is uniform. With these relations and Equation (I.13), we have

$$\frac{1}{p_1}\left(\frac{\partial p}{\partial b}\right)_2 = \frac{4}{\gamma+1}\frac{w\sin^2\beta\,\tan\beta}{F_{x_1}^4}\left[F_{x_1}\,F_{x_2}\,F_{x_3}\left(F_{x_2x_2}-F_{x_3x_3}\right)\right.$$

$$\left. - \psi F_{x_3}F_{x_1x_2} + \psi F_{x_2}F_{x_1x_3} + F_{x_1}\left(F_{x_3}^2 - F_{x_2}^2\right)F_{x_2x_3}\right] \qquad \text{(G.21b)}$$

With the further simplification of an elliptic paraboloid EP shock (Appendix J), we obtain

$$\frac{1}{p_1}\left(\frac{\partial p}{\partial b}\right)_2 = \frac{4}{\gamma+1}\,w\sin^2\beta\,\tan\beta\left[\frac{\sigma x_2 x_3}{r_2 r_3}\left(\frac{1}{r_3}-\frac{1}{r_2}\right)\right]$$

$$= \frac{4\gamma}{\gamma+1}\frac{m}{\left(1+\psi\right)^2\psi^{1/2}}\frac{\sigma x_2 x_3}{r_2 r_3}\left(\frac{1}{r_3}-\frac{1}{r_2}\right) \qquad \text{(G.21c)}$$

Note that this is zero only when one of the following conditions is satisfied:

$$x_2 = 0, \qquad x_3 = 0, \qquad \sigma = 0, \qquad r_2 = r_3 \qquad \text{(G.22)}$$

If the EP shock is not two-dimensional or axisymmetric and the point of interest is not on a $x_2 = 0$ or $x_3 = 0$ symmetry curve, there is a finite pressure gradient along b whose sign depends on the sign of $x_2 x_3 (r_2 - r_3)$.

Appendix H: Steady Shock Derivative Formulation

H.1   COMMON ITEMS

$$F = F(x_i) = 0 \tag{H.1}$$

$$\hat{n} - \frac{\nabla F}{|\nabla F|}, \qquad \nabla F = \sum F_{x_j}\, \hat{1}_j, \qquad |\nabla F| = \left(\sum F_{x_j}^2\right)^{1/2} \tag{H.2}$$

$$\sin\beta = \frac{\sum v_{1,j}F_{x_j}}{V_1\,|\nabla F|}, \qquad \cos\beta = \frac{\left(\sum K_j^2\right)^{1/2}}{V_1\,|\nabla F|} \tag{H.3}$$

$$\chi = \frac{1}{V_1\,|\nabla F|\cos\beta} = \frac{1}{\left(\sum K_j^2\right)^{1/2}} \tag{H.4}$$

$$\hat{b} = -\chi\sum K_j\, \hat{1}_j \tag{H.5}$$

$$K_1 = v_{1,3}F_{x_2} - v_{1,2}F_{x_3}, \qquad K_2 = v_{1,1}F_{x_3} - v_{1,3}F_{x_1}, \qquad K_3 = v_{1,2}F_{x_1} - v_{1,1}F_{x_2} \tag{H.6}$$

$$\hat{t} = \frac{\chi}{|\nabla F|}\sum L_j\, \hat{1}_j \tag{H.7}$$

$$L_1 = F_{x_3}K_2 - F_{x_2}K_3, \qquad L_2 = F_{x_1}K_3 - F_{x_3}K_1, \qquad L_3 = F_{x_2}K_1 - F_{x_1}K_2 \tag{H.8}$$

$$\sum F_{x_j}K_j = \sum F_{x_j}L_j = \sum K_j L_j = \sum v_{1,j}K_j = 0 \tag{H.9}$$

$$\sum K_j^2 = \frac{1}{\chi^2}, \qquad \sum L_j^2 = \frac{|\nabla F|^2}{\chi^2} \tag{H.10}$$

$$\vec{V}_2 = V_2\hat{s}_2, \qquad V_2 = \frac{2}{\gamma+1}V_1\frac{VB^{1/2}}{(mw)^{1/2}} \tag{H.11}$$

$$\hat{s}_2 = \frac{1}{\cos\beta}\sum\left[-\frac{\sin\theta}{|\nabla F|}F_{x_j} + \frac{\cos(\beta-\theta)}{V_1}v_{j,1}\right]\hat{1}_j = \sin(\beta-\theta)\hat{n} + \cos(\beta-\theta)\hat{t} \tag{H.12}$$

H.2 S_a, S_b

For S_a use Equations (9.21a), (9.22a), and (9.26). For S_b use Equations (9.23b), (9.25b), and (9.26).

H.3 $\vec{\omega}_2$

Use Equations (9.48) through (9.51b), where M_{1x_i} and β_{x_i} are given by Equations (9.33) and (9.36), respectively.

H.4   TANGENTIAL DERIVATIVES

Equations for β_{x_i}, M_1, M_{x_i}, $\partial()/\partial s$, $\partial()/\partial b$, $\partial w/\partial s$, and $\partial w/\partial b$ are given by Equations (9.35), (9.32), (9.33), (G.17a,b), and (9.54a,b), respectively.

$$\frac{1}{V_1}\left(\frac{\partial u}{\partial s}\right)_2 = \frac{\chi}{|\nabla F|}\left(\frac{\cos\beta}{V_1^2}\sum_j v_{1,j}\sum_i L_i v_{1,jx_i} - \sin\beta\sum L_i\beta_{x_i}\right) \tag{H.13}$$

$$\frac{1}{V_1}\left(\frac{\partial u}{\partial b}\right)_2 = -\chi\left(\frac{\cos\beta}{V_1^2}\sum_j v_{1,j}\sum_i K_i v_{1,jx_i} - \sin\beta\sum K_i\beta_{x_i}\right) \tag{H.14}$$

$$\frac{1}{V_1}\left(\frac{\partial v}{\partial s}\right)_2 = \frac{2}{\gamma+1}\frac{1}{w}\frac{\chi}{|\nabla F|}\left[\frac{X\sin\beta}{V_1^2}\sum_j v_{1,j}\sum_i L_i v_{1,jx_i} - 2\sin\beta\sum L_i\frac{M_{1x_i}}{M_1}\right.$$
$$\left. -\left(1-\frac{\gamma-1}{2}w\right)\cos\beta\sum L_i\beta_{x_i}\right] \tag{H.15}$$

$$\frac{1}{V_1}\left(\frac{\partial v}{\partial b}\right)_2 = -\frac{2}{\gamma+1}\frac{\chi}{w}\left[\frac{X\sin\beta}{V_1^2}\sum_j v_{1,j}\sum_i K_i v_{1,jx_i} - 2\sin\beta\sum K_i\frac{M_{1x_i}}{M_1}\right.$$
$$\left. -\left(1-\frac{\gamma-1}{2}w\right)\cos\beta\sum K_i\beta_{x_i}\right] \tag{H.16}$$

$$\frac{1}{p_1}\left(\frac{\partial p}{\partial s}\right)_2 = \frac{2}{\gamma+1}\frac{\chi}{|\nabla F|}\left[Y\sum L_i\frac{p_{1x_i}}{p_1} + 2\gamma w\sum L_i\frac{M_{1x_i}}{M_1} + 2\gamma w\cot\beta\sum L_i\beta_{x_i}\right] \tag{H.17}$$

$$\frac{1}{p_1}\left(\frac{\partial p}{\partial b}\right)_2 = -\frac{2}{\gamma+1}\chi\left[Y\sum K_i\frac{p_{1x_i}}{p_1} + 2\gamma w\sum K_i\frac{M_{1x_i}}{M_1} + 2\gamma w\cot\beta\sum K_i\beta_{x_i}\right] \tag{H.18}$$

$$\frac{1}{\rho_1}\left(\frac{\partial \rho}{\partial s}\right)_2 = \frac{\gamma+1}{2}\frac{w}{X^2}\frac{\chi}{|\nabla F|}\left(X\sum L_i\frac{\rho_{1x_i}}{\rho_1} + 2\sum L_i\frac{M_{1x_i}}{M_1} + 2\cot\beta\sum L_i\beta_{x_i}\right)$$

$$(H.19)$$

$$\frac{1}{\rho_1}\left(\frac{\partial \rho}{\partial b}\right)_2 = -\frac{\gamma+1}{2}\frac{w\chi}{X^2}\left(X\sum K_i\frac{\rho_{1x_i}}{\rho_1} + 2\sum K_i\frac{M_{1x_i}}{M_1} + 2\cot\beta\sum K_i\beta_{x_i}\right) \quad (H.20)$$

H.5　NORMAL DERIVATIVES

$$\vec{V} = u\hat{t} + v\hat{n} + \bar{w}\hat{b} \tag{H.21}$$

$$\nabla = \hat{t}\frac{\partial}{\partial s} + \hat{n}\frac{\partial}{\partial b} + \hat{b}\frac{\partial}{\partial b} \tag{H.22}$$

$$\bar{w} = 0,\qquad \frac{\partial \bar{w}}{\partial s} = \frac{\partial \bar{w}}{\partial b} = 0,\qquad \frac{\partial \bar{w}}{\partial n} \neq 0 \tag{H.23}$$

$$\frac{1}{h_1}\frac{\partial h_1}{\partial b} = -\frac{\chi^3}{|\nabla F|^2}\sum_i L_i \sum_j L_j \frac{\partial K_i}{\partial x_j} \tag{H.24}$$

$$\frac{1}{h_2}\frac{\partial h_2}{\partial b} = -\frac{\chi}{|\nabla F|^2}\sum_i F_{x_i} \sum_j F_{x_j} \frac{\partial K_i}{\partial x_j} \tag{H.25}$$

$$\frac{1}{h_3}\frac{\partial h_3}{\partial s} = -\frac{\chi^3}{|\nabla F|}\sum_i L_i \sum_j K_j \frac{\partial K_i}{\partial x_j} \tag{H.26}$$

$$\left(\frac{1}{h_2}\frac{\partial h_2}{\partial s}\right)_{surface} = 0 \tag{H.27}$$

$$A_a = -\frac{\partial u}{\partial s} - \frac{u}{\rho}\frac{\partial \rho}{\partial s} + v\left(S_a + S_b\right) - \frac{u}{h_3}\frac{\partial h_3}{\partial s} \tag{H.28}$$

$$A_c = -u\frac{\partial v}{\partial s} - u^2 S_a \tag{H.29}$$

$$A_e = -u\frac{\partial p}{\partial s} + \frac{\gamma p u}{\rho}\frac{\partial \rho}{\partial s} \tag{H.30}$$

$$\frac{\partial u}{\partial n} = \frac{1}{V_1}\left(\frac{\partial u}{\partial n}\right)_2 = \frac{1}{v}\left(-u\frac{\partial u}{\partial s} + uv\,S_a - \frac{1}{\gamma m\rho}\frac{\partial p}{\partial s}\right) \tag{H.31}$$

$$\frac{\partial v}{\partial n} = \frac{1}{V_1}\left(\frac{\partial v}{\partial n}\right)_2 = \frac{1}{\dfrac{p}{m\rho} - v^2}\left(\frac{p}{m\rho}A_a - vA_c + \frac{1}{\gamma m\rho}A_e\right) \qquad \text{(H.32)}$$

$$\frac{\partial p}{\partial n} = \frac{1}{p_1}\left(\frac{\partial p}{\partial n}\right)_2 = \frac{1}{\dfrac{p}{m\rho} - v^2}\left(-\gamma v p A_a + \gamma p A_c - vA_e\right) \qquad \text{(H.33)}$$

$$\frac{\partial \rho}{\partial n} = \frac{1}{\rho_1}\left(\frac{\partial \rho}{\partial n}\right)_2 = \frac{1}{\dfrac{p}{m\rho} - v^2}\left(-\rho v A_a + \rho A_c - \frac{1}{\gamma m v}A_e\right) \qquad \text{(H.34)}$$

$$\frac{\partial \bar{w}}{\partial n} = \frac{1}{V_1}\left(\frac{\partial \bar{w}}{\partial n}\right)_2 = \frac{1}{v}\left(\frac{u^2}{h_1}\frac{\partial h_1}{\partial b} + \frac{v^2}{h_2}\frac{\partial h_2}{\partial b} - \frac{1}{\gamma m\rho}\frac{\partial p}{\partial b}\right) \qquad \text{(H.35)}$$

Appendix I: Uniform Freestream Formulation

I.1 COMMON ITEMS

$$v_{1,1} = V_1 = \text{constant}, \quad v_{1,2} = v_{1,3} = 0, \quad M_{x_i} = 0, \quad \omega_1 = 0 \tag{I.1}$$

$$\psi = F_{x_2}^2 + F_{x_3}^2 \tag{I.2}$$

$$|\nabla F| = \left(F_{x_1}^2 + \psi\right)^{1/2} \tag{I.3}$$

$$\chi = \frac{1}{V_1 \psi^{1/2}} \tag{I.4}$$

$$\sin\beta = \frac{F_{x_1}}{|\nabla F|}, \quad \cos\beta = \frac{\psi^{1/2}}{|\nabla F|}, \quad \tan\theta = \frac{1}{\tan\beta} \frac{M_1^2 \sin^2\beta - 1}{1 + \left(\dfrac{\gamma+1}{2} - \sin^2\beta\right)M_1^2} \tag{I.5}$$

$$\beta_{x_i} = \frac{1}{\psi^{1/2}|\nabla F|^2}\left[\psi F_{x_1 x_i} - F_{x_1}\left(F_{x_2}F_{x_2 x_i} + F_{x_3}F_{x_3 x_i}\right)\right] \tag{I.6}$$

$$K_1 = 0, \quad K_2 = V_1 F_{x_3}, \quad K_3 = -V_1 F_{x_2} \tag{I.7}$$

$$L_1 = V_1\psi, \quad L_2 = -V_1 F_{x_1} F_{x_2}, \quad L_3 = -V_1 F_{x_1} F_{x_3} \tag{I.8}$$

$$\hat{t} = \frac{1}{|\nabla F|\psi^{1/2}}\left[\psi\,\hat{1}_1 - F_{x_1}\left(F_{x_2}\,\hat{1}_2 + F_{x_3}\,\hat{1}_3\right)\right] \tag{I.9}$$

$$\hat{n} = \frac{1}{|\nabla F|}\sum F_{x_j}\,\hat{1}_j \tag{I.10}$$

$$\hat{b} = -\frac{1}{\psi^{1/2}}\left(F_{x_3}\,\hat{1}_2 - F_{x_2}\,\hat{1}_3\right) \tag{I.11}$$

I.2 S_a, S_b

Use Appendix H.2, but with L_j^* and K_j^* replaced with Equations (I.8) and (I.7), respectively.

I.3 $\vec{\omega}_2$

$$\sum L_i \beta_{x_i} = \frac{V_1}{\psi^{1/2} \, |\nabla F|^2} \left[\psi^2 F_{x_1 x_1} - 2\psi F_{x_1} \left(F_{x_2} F_{x_1 x_2} + F_{x_3} F_{x_1 x_3} \right) \right.$$

$$\left. + F_{x_1}^2 \left(F_{x_2}^2 F_{x_2 x_2} + 2 F_{x_2} F_{x_3} F_{x_2 x_3} + F_{x_3}^2 F_{x_3 x_3} \right) \right] \tag{I.12}$$

$$\sum K_i \beta_{x_i} = - \frac{V_1}{\psi^{1/2} \, |\nabla F|^2} \left[F_{x_1} F_{x_2} F_{x_3} \left(F_{x_2 x_2} - F_{x_3 x_3} \right) - \psi F_{x_3} F_{x_1 x_2} \right.$$

$$\left. + \psi F_{x_2} F_{x_1 x_3} + F_{x_1} \left(F_{x_3}^2 - F_{x_2}^2 \right) F_{x_2 x_3} \right] \tag{I.13}$$

$$Q_b = \frac{2}{\gamma + 1} \frac{Z^2}{wX} \frac{1}{|\nabla F|^2} \sum L_i \beta_{x_i} \tag{I.14}$$

$$Q_t = \frac{2}{\gamma + 1} \frac{Z^2}{wX} \frac{1}{|\nabla F|} \sum K_i \beta_{x_i} \tag{I.15}$$

$$\vec{\omega}_2 = -Q_b \hat{b} - Q_t \hat{t} \tag{I.16}$$

$$\omega_2 = \pm \frac{2}{\gamma + 1} \frac{Z^2}{wX} \frac{1}{|\nabla F|^2} \left[\left(\sum L_i b_{x_i} \right)^2 + |\nabla F| \left(\sum K_i b_{x_i} \right)^2 \right]^{1/2} \tag{I.17}$$

I.4 TANGENTIAL DERIVATIVES

$$\frac{\partial}{\partial s} = \frac{1}{\psi^{1/2} \, |\nabla F|} \left(\psi \frac{\partial}{\partial x_1} - F_{x_1} F_{x_2} \frac{\partial}{\partial x_2} - F_{x_1} F_{x_3} \frac{\partial}{\partial x_3} \right) \tag{I.18}$$

$$\frac{\partial}{\partial b} = - \frac{1}{\psi^{1/2}} \left(F_{x_3} \frac{\partial}{\partial x_2} - F_{x_2} \frac{\partial}{\partial x_3} \right) \tag{I.19}$$

$$\frac{1}{w} \frac{\partial w}{\partial s} = 2 \cot \beta \, \frac{\chi}{|\nabla F|} \sum L_i \beta_{x_i} \tag{I.20}$$

$$\frac{1}{w} \frac{\partial w}{\partial b} = -2 \cot \beta \, \chi \sum K_i \beta_{x_i} \tag{I.21}$$

$$\frac{1}{V_1} \left(\frac{\partial u}{\partial s} \right)_2 = - \frac{\chi \sin \beta}{|\nabla F|} \sum L_i \beta_{x_i} \tag{I.22}$$

$$\frac{1}{V_1}\left(\frac{\partial u}{\partial b}\right)_2 = -\chi\sin\beta \sum K_i\beta_{x_i} \qquad (I.23)$$

$$\frac{1}{V_1}\left(\frac{\partial v}{\partial s}\right)_2 = \frac{2}{\gamma+1}\frac{1-\dfrac{\gamma-1}{2}w}{w}\frac{\chi}{|\nabla F|}\cos\beta \sum L_i\beta_{x_i} \qquad (I.24)$$

$$\frac{1}{V_1}\left(\frac{\partial v}{\partial b}\right)_2 = \frac{2}{\gamma+1}\frac{1-\dfrac{\gamma-1}{2}w}{w}\chi\cos\beta \sum K_i\beta_{x_i} \qquad (I.25)$$

$$\frac{1}{p_1}\left(\frac{\partial p}{\partial s}\right)_2 = \frac{4\gamma}{\gamma+1}w\cot\beta\,\frac{\chi}{|\nabla F|} \sum L_i\beta_{x_i} \qquad (I.26)$$

$$\frac{1}{p_1}\left(\frac{\partial p}{\partial b}\right)_2 = -\frac{4\gamma}{\gamma+1}w\cot\beta\,\chi \sum K_i\beta_{x_i} \qquad (I.27)$$

$$\frac{1}{\rho_1}\left(\frac{\partial\rho}{\partial s}\right)_2 = (\gamma+1)\frac{w}{X^2}\frac{\chi}{|\nabla F|}\cot\beta \sum L_i\beta_{x_i} \qquad (I.28)$$

$$\frac{1}{\rho_1}\left(\frac{\partial\rho}{\partial b}\right)_2 = -(\gamma+1)\frac{w}{X^2}\chi\cot\beta \sum K_i\beta_{x_i} \qquad (I.29)$$

I.5 NORMAL DERIVATIVES

$$\frac{1}{h_1}\frac{\partial h_1}{\partial b} = -\frac{F_{x_1}}{|\nabla F|^2\,\psi^{3/2}}\Big\{\psi\left(F_{x_3}F_{x_1x_2} - F_{x_2}F_{x_1x_3}\right)$$

$$+ F_{x_1}\Big[-F_{x_2}F_{x_3}F_{x_2x_2} + \left(F_{x_2}^2 - F_{x_3}^2\right)F_{x_2x_3} + F_{x_2}F_{x_3}F_{x_3x_3}\Big]\Big\} \qquad (I.30)$$

$$\frac{1}{h_2}\frac{\partial h_2}{\partial b} = \frac{1}{|\nabla F|^2\,\psi^{1/2}}\Big[-F_{x_2}\left(F_{x_1}F_{x_1x_3} + F_{x_2}F_{x_2x_3} + F_{x_3}F_{x_3x_3}\right)$$

$$+ F_{x_3}\left(F_{x_1}F_{x_1x_2} + F_{x_2}F_{x_2x_2} + F_{x_3}F_{x_2x_3}\right)\Big] \qquad (I.31)$$

$$\frac{1}{h_3}\frac{\partial h_3}{\partial s} = -\frac{1}{|\nabla F|\,\psi^{3/2}}\left(F_{x_2}^2 F_{x_3x_3} - 2\,F_{x_2}F_{x_3}F_{x_2x_3} + F_{x_3}^2 F_{x_2x_2}\right) \qquad (I.32)$$

For Equations (H.28) through (H.35), use Appendix E for u, v, p, and ρ and Appendices I.2 and I.4 for the tangential derivatives and S_a and S_b.

Appendix J: Elliptic Paraboloid Shock Formulation

J.1 COMMON ITEMS

$$F = x_1 - \frac{x_2^2}{2r_2} - \frac{\sigma x_3^2}{2r_3} = 0 \tag{J.1a}$$

$$F_{x_1} = 1, \quad F_{x_2} = -\frac{x_2}{r_2}, \quad F_{x_3} = -\frac{\sigma x_3}{r_3} \tag{J.1b}$$

$$F_{x_1x_1} = 0, \quad F_{x_1x_2} = 0, \quad F_{x_1x_3} = 0, \quad F_{x_2x_3} = 0, \quad F_{x_2x_2} = -\frac{1}{r_2}, \quad F_{x_3x_3} = -\frac{\sigma}{r_3} \tag{J.2}$$

$$\psi = \left(\frac{x_2}{r_2}\right)^2 + \left(\frac{\sigma x_3}{r_3}\right)^2, \quad \chi = \frac{1}{V_1 \psi^{1/2}}, \quad |\nabla F| = (1+\psi)^{1/2} \tag{J.3}$$

$$\sin\beta = \frac{1}{(1+\psi)^{1/2}}, \quad \cos\beta = \frac{\psi^{1/2}}{(1+\psi)^{1/2}} \tag{J.4}$$

$$\beta_{x_1} = 0, \quad \beta_{x_2} = -\frac{1}{\psi^{1/2}(1+\psi)}\frac{x_2}{r_2^2}, \quad \beta_{x_3} = -\frac{1}{\psi^{1/2}(1+\psi)}\left(\frac{\sigma x_3}{r_3^2}\right) \tag{J.5}$$

$$K_1 = 0, \quad K_2 = -V_1 \frac{\sigma x_3}{r_3}, \quad K_3 = V_1 \frac{x_2}{r_2} \tag{J.6}$$

$$L_1 = V_1\psi, \quad L_2 = V_1\frac{x_2}{r_2}, \quad L_3 = V_1\frac{\sigma x_3}{r_3} \tag{J.7}$$

$$\hat{t} = \frac{1}{(1+\psi)^{1/2}\psi^{1/2}}\left(\psi\,\hat{i}_1 + \frac{x_2}{r_2}\,\hat{i}_2 + \frac{\sigma x_3}{r_3}\,\hat{i}_3\right) \tag{J.8}$$

$$\hat{n} = \frac{1}{(1+\psi)^{1/2}}\left(\hat{i}_1 - \frac{x_2}{r_2}\,\hat{i}_2 - \frac{\sigma x_3}{r_3}\,\hat{i}_3\right) \tag{J.9}$$

$$\hat{b} = \frac{1}{\psi^{1/2}}\left(\frac{\sigma x_3}{r_3}\,\hat{i}_2 - \frac{x_2}{r_2}\,\hat{i}_3\right) \tag{J.10}$$

J.2 S_a, S_b

$$S_a = \frac{1}{\psi(1+\psi)^{3/2}} \left(\frac{x_2^2}{r_2^3} + \frac{\sigma x_3^2}{r_3^3} \right) \tag{J.11}$$

$$S_b = \frac{1}{\psi(1+\psi)^{1/2}} \frac{\sigma}{r_2 r_3} \left(\frac{x_2^2}{r_2} + \frac{x_3^2}{r_3} \right) \tag{J.12}$$

J.3 $\vec{\omega}_2$

$$\sum L_i \beta_{x_i} = - \frac{V_1}{\psi^{1/2}(1+\psi)} \left(\frac{x_2^2}{r_2^3} + \frac{\sigma x_3^2}{r_3^3} \right) \tag{J.13}$$

$$\sum K_i \beta_{x_i} = \frac{\sigma V_1}{\psi^{1/2}(1+\psi)} \left(\frac{1}{r_2} - \frac{1}{r_3} \right) \frac{x_2}{r_2} \frac{x_3}{r_3} \tag{J.14}$$

$$Q_b = - \frac{2}{\gamma+1} V_1 \frac{1}{(1+\psi)^2 \psi^{1/2}} \frac{Z^2}{wX} \left(\frac{x_2^2}{r_2^3} + \frac{\sigma x_3^2}{r_3^3} \right) \tag{J.15}$$

$$Q_t = \frac{2}{\gamma+1} \sigma V_1 \frac{1}{(1+\psi)^{3/2} \psi^{1/2}} \frac{Z^2}{wX} \left(\frac{1}{r_2} - \frac{1}{r_3} \right) \frac{x_2}{r_2} \frac{x_3}{r_3} \tag{J.16}$$

$$\vec{\omega}_2 = -Q_b \hat{b} - Q_t \hat{t} \tag{J.17}$$

J.4   TANGENTIAL DERIVATIVES

$$\frac{\partial}{\partial s} = \frac{1}{\psi^{1/2}(1+\psi)^{1/2}} \left(\psi \frac{\partial}{\partial x_1} + \frac{x_2}{r_2} \frac{\partial}{\partial x_2} + \frac{\sigma x_3}{r_3} \frac{\partial}{\partial x_3} \right) \tag{J.18}$$

$$\frac{\partial}{\partial b} = - \frac{1}{\psi^{1/2}} \left(- \frac{\sigma x_3}{r_3} \frac{\partial}{\partial x_2} + \frac{x_2}{r_2} \frac{\partial}{\partial x_3} \right) \tag{J.19}$$

$$\frac{1}{w} \frac{\partial w}{\partial s} = - \frac{2}{(1+\psi)^{3/2} \psi^{1/2}} \left(\frac{x_2^2}{r_2^3} + \frac{\sigma x_3^2}{r_3^3} \right) \tag{J.20}$$

$$\frac{1}{w} \frac{\partial w}{\partial b} = - \frac{2}{(1+\psi)\psi^{1/2}} \frac{\sigma x_2 x_3}{r_2 r_3} \left(\frac{1}{r_2} - \frac{1}{r_3} \right) \tag{J.21}$$

$$\frac{1}{V_1} \left(\frac{\partial u}{\partial s} \right)_2 = \frac{1}{(1+\psi)^2 \psi} \left(\frac{x_2^2}{r_2^3} + \frac{\sigma x_3^2}{r_3^3} \right) \tag{J.22}$$

$$\frac{1}{V_1} \left(\frac{\partial u}{\partial b} \right)_2 = \frac{1}{(1+\psi)^{3/2} \psi} \frac{\sigma x_2 x_3}{r_2 r_3} \left(\frac{1}{r_2} - \frac{1}{r_3} \right) \tag{J.23}$$

$$\frac{1}{V_1}\left(\frac{\partial v}{\partial s}\right)_2 = \frac{2}{\gamma+1}\frac{1-\frac{\gamma-1}{2}w}{w}\frac{1}{(1+\psi)^2\psi^{1/2}}\left(\frac{x_2^2}{r_2^3}+\frac{\sigma x_3^2}{r_3^3}\right) \tag{J.24}$$

$$\frac{1}{V_1}\left(\frac{\partial v}{\partial b}\right)_2 = \frac{2}{\gamma+1}\frac{1-\frac{\gamma-1}{2}w}{w}\frac{1}{(1+\psi)^{3/2}\psi^{1/2}}\frac{\sigma x_2 x_3}{r_2 r_3}\left(\frac{1}{r_2}-\frac{1}{r_3}\right) \tag{J.25}$$

$$\frac{1}{p_1}\left(\frac{\partial p}{\partial s}\right)_2 = -\frac{4\gamma}{\gamma+1}\frac{m}{(1+\psi)^{5/2}\psi^{1/2}}\left(\frac{x_2^2}{r_2^3}+\frac{\sigma x_3^2}{r_3^3}\right) \tag{J.26}$$

$$\frac{1}{p_1}\left(\frac{\partial p}{\partial b}\right)_2 = -\frac{4\gamma}{\gamma+1}\frac{m}{(1+\psi)^2\psi^{1/2}}\frac{\sigma x_2 x_3}{r_2 r_3}\left(\frac{1}{r_2}-\frac{1}{r_3}\right) \tag{J.27}$$

$$\frac{1}{\rho_1}\left(\frac{\partial\rho}{\partial s}\right)_2 = -(\gamma+1)\frac{m}{(1+\psi)^{5/2}\psi^{1/2}X^2}\left(\frac{x_2^2}{r_2^3}+\frac{\sigma x_3^2}{r_3^3}\right) \tag{J.28}$$

$$\frac{1}{\rho_1}\left(\frac{\partial\rho}{\partial b}\right)_2 = -(\gamma+1)\frac{m}{(1+\psi)^2\psi^{1/2}X^2}\frac{\sigma x_2 x_3}{r_2 r_3}\left(\frac{1}{r_2}-\frac{1}{r_3}\right) \tag{J.29}$$

J.5   NORMAL DERIVATIVES

$$\frac{1}{h_1}\frac{\partial h_1}{\partial b} = \frac{1}{(1+\psi)\psi^{3/2}}\frac{\sigma x_2 x_3}{r_2 r_3}\left(\frac{1}{r_3}-\frac{1}{r_2}\right) \tag{J.30}$$

$$\frac{1}{h_2}\frac{\partial h_2}{\partial b} = \frac{1}{(1+\psi)\psi^{1/2}}\frac{\sigma x_2 x_3}{r_2 r_3}\left(\frac{1}{r_3}-\frac{1}{r_2}\right) \tag{J.31}$$

$$\frac{1}{h_3}\frac{\partial h_3}{\partial s} = \frac{1}{(1+\psi)^{1/2}\psi^{3/2}}\frac{\sigma}{r_2 r_3}\left(\frac{x_2^2}{r_2}+\frac{x_3^2}{r_3}\right) \tag{J.32}$$

$$w = \frac{m}{1+\psi} \tag{J.33}$$

$$u = \frac{u_2}{V_1} = \frac{\psi^{1/2}}{(1+\psi)^{1/2}} \tag{J.34}$$

$$v = \frac{v_2}{V_1} = \frac{2}{\gamma+1}\frac{X}{m}(1+\psi)^{1/2} \tag{J.35}$$

$$p = \frac{p_2}{p_1} = \frac{2}{\gamma+1} Y \tag{J.36}$$

$$\rho = \frac{\rho_2}{\rho_1} = \frac{\gamma+1}{2} \frac{m}{X} \frac{1}{1+\psi} \tag{J.37}$$

$$A_a = \frac{2}{\psi(1+\psi)^2} \left(\frac{\psi}{X} - \frac{1}{\gamma+1} \frac{Z}{w} \right) \left(\frac{x_2^2}{r_2^3} + \frac{\sigma x_3^2}{r_3^3} \right) - \frac{2}{\gamma+1} \frac{1}{\psi(1+\psi)} \frac{Z}{w} \frac{\sigma}{r_2 r_3} \left(\frac{x_2^2}{r_2} + \frac{x_3^2}{r_3} \right) \tag{J.38}$$

$$A_c = -\frac{2}{\gamma+1} \frac{1}{(1+\psi)^{5/2}} \frac{1+w}{w} \left(\frac{x_2^2}{r_2^3} + \frac{\sigma x_3^2}{r_3^3} \right) \tag{J.39}$$

$$A_e = \frac{2\gamma(\gamma-1)}{(\gamma+1)} \frac{1}{(1+\psi)^2} \frac{Z^2}{X} \left(\frac{x_2^2}{r_2^3} + \frac{\sigma x_3^2}{r_3^3} \right) \tag{J.40}$$

$$\frac{\partial u}{\partial n} = \frac{1}{V_1} \left(\frac{\partial u}{\partial n} \right)_2 = \frac{1}{(1+\psi)^3 \psi^{1/2}} \left[-\frac{\gamma+1}{2} \frac{m}{X} + \frac{\gamma+5}{\gamma+1} (1+\psi) \right] \left(\frac{x_2^2}{r_2^3} + \frac{\sigma x_3^2}{r_3^3} \right) \tag{J.41}$$

$$\frac{\partial v}{\partial n} = \frac{1}{V_1} \left(\frac{\partial v}{\partial n} \right)_2 = \frac{1}{Z} \left[\frac{2}{\gamma+1} Y A_a - \frac{m}{(1+\psi)^{1/2}} A_c + \frac{1}{\gamma} A_e \right] \tag{J.42}$$

$$\frac{\partial p}{\partial n} = \frac{1}{p_1} \left(\frac{\partial p}{\partial n} \right)_2 = \frac{m}{(1+\psi)^{1/2} Z} \left[-\frac{2\gamma}{\gamma+1} Y A_a + \frac{\gamma m Y}{X} \frac{1}{(1+\psi)^{1/2}} A_c - A_e \right] \tag{J.43}$$

$$\frac{\partial \rho}{\partial n} = \frac{1}{\rho_1} \left(\frac{\partial \rho}{\partial n} \right)_2 = \frac{\gamma+1}{2} \frac{m^2}{X^2 Z} \frac{1}{(1+\psi)^2} \left[-X (1+\psi)^{1/2} A_a + \frac{\gamma+1}{2} m A_c \right.$$

$$\left. - \frac{\gamma+1}{2\gamma} (1+\psi)^{1/2} A_e \right] \tag{J.44}$$

$$\frac{\partial \bar{w}}{\partial n} = \frac{1}{V_1} \left(\frac{\partial \bar{w}}{\partial n} \right)_2 = \frac{\gamma+1}{2} \frac{1}{(1+\psi)^{5/2} \psi^{1/2}} \left\{ \left[\frac{m}{X} + \left(\frac{2}{\gamma+1} \right)^2 \frac{X}{m} (1+\psi)^{5/2} \right] \right.$$

$$\left. \times \frac{\sigma x_2 x_3}{r_2 r_3} \left(\frac{1}{r_3} - \frac{1}{r_2} \right) - \frac{2}{\gamma(\gamma+1)} \frac{(1+\psi)^3 \psi^{1/2}}{m} \frac{1}{p_1} \frac{\partial p}{\partial b} \right\} \tag{J.45}$$

When the shock is two-dimensional or axisymmetric, set $\partial \bar{w}/\partial n = 0$.

Appendix K: Global, Shock-Based Coordinates

By means of examples, this appendix demonstrates the nonexistence of shock-based coordinates, ξ_i, for a three-dimensional shock. A variety of approaches were tried. The one discussed is perhaps the easiest to follow; it is similar to the analysis in Chapter 3.

Coordinates are assumed such that

$$\nabla \xi_1 = \frac{1}{h_1}\,\hat{t}, \qquad \nabla \xi_2 = \frac{1}{h_2}\,\hat{n}, \qquad \nabla \xi_3 = \frac{1}{h_3}\,\hat{b} \tag{K.1a}$$

where the h_i are scale factors and

$$\xi_1 \sim s, \quad \xi_2 \sim n, \quad \xi_3 \sim b \tag{K.2}$$

With $\hat{t}, \hat{n}, \hat{b}$ given by Equations (9.10), (9.2), and (9.7b), respectively, we obtain

$$\nabla \xi_1 = \sum \frac{\partial \xi_1}{\partial x_j}\,\hat{\imath}_j = \frac{\chi}{h_1\,|\nabla F|} \sum L_j\,\hat{\imath}_j \tag{K.1b}$$

$$\nabla \xi_2 = \sum \frac{\partial \xi_2}{\partial x_j}\,\hat{\imath}_j = \frac{1}{h_2\,|\nabla F|} \sum F_{x_j}\,\hat{\imath}_j \tag{K.1c}$$

$$\nabla \xi_3 = \sum \frac{\partial \xi_3}{\partial x_j}\,\hat{\imath}_j = -\frac{\chi}{h_3} \sum K_j\,\hat{\imath}_j \tag{K.1d}$$

This yields the array

$$\frac{\partial \xi_1}{\partial x_i} = \frac{\chi L_i}{h_1\,|\nabla F|}, \qquad \frac{\partial \xi_2}{\partial x_i} = \frac{F_{x_i}}{h_2\,|\nabla F|}, \qquad \frac{\partial \xi_3}{\partial x_i} = -\frac{\chi K_i}{h_3}, \qquad i = 1, 2, 3 \tag{K.3}$$

which is in accord with Appendix G. The scale factors are evaluated using Equation (3.19), the existence condition. For instance, for ξ_2 we write

$$\frac{\partial^2 \xi_2}{\partial x_2\,\partial x_1} = \frac{\partial}{\partial x_2}\left(\frac{F_{x_1}}{h_2\,|\nabla F|}\right) = \frac{F_{x_1 x_2}}{h_2\,|\nabla F|} - \frac{F_{x_1}}{|\nabla F|}\frac{1}{h_2^2}\frac{\partial h_2}{\partial x_2} - \frac{F_{x_1}}{h_2\,|\nabla F|^2}\frac{\partial |\nabla F|}{\partial x_2}$$

and a similar result for $\partial^2 \xi_2 / (\partial x_1\,\partial x_2)$. Equating the two equations then yields

$$F_{x_2}\frac{\partial q_2}{\partial x_1} - F_{x_1}\frac{\partial q_2}{\partial x_2} + 0\frac{\partial q_2}{\partial x_3} = \frac{F_{x_1}}{|\nabla F|}\frac{\partial |\nabla F|}{\partial x_2} - \frac{F_{x_2}}{|\nabla F|}\frac{\partial |\nabla F|}{\partial x_1} \tag{K.4}$$

where

$$q_2 = \ln h_2$$

More generally, the three partial differential equations (PDEs) for q_2 are

$$F_{x2}\frac{\partial q_2^{(a)}}{\partial x_1} - F_{x1}\frac{\partial q_2^{(a)}}{\partial x_2} + 0\,\frac{\partial q_2^{(a)}}{\partial x_3} + \frac{1}{|\nabla F|^2}\left(F_{x2}\sum F_{xj}F_{xjx1} - F_{x1}\sum F_{xj}F_{xjx2} \right) = 0$$

(K.5a)

$$F_{x3}\frac{\partial q_2^{(b)}}{\partial x_1} + 0\,\frac{\partial q_2^{(b)}}{\partial x_2} - F_{x1}\frac{\partial q_2^{(b)}}{\partial x_3} + \frac{1}{|\nabla F|^2}\left(F_{x3}\sum F_{xj}F_{xjx1} - F_{x1}\sum F_{xj}F_{xjx3} \right) = 0$$

(K.5b)

$$0\,\frac{\partial q_2^{(c)}}{\partial x_1} + F_{x3}\frac{\partial q_2^{(c)}}{\partial x_2} - F_{x2}\frac{\partial q_2^{(c)}}{\partial x_3} + \frac{1}{|\nabla F|^2}\left(F_{x3}\sum F_{xj}F_{xjx2} - F_{x2}\sum F_{xj}F_{xjx3} \right) = 0$$

(K.5c)

As is done in Chapter 3, these equations are solved by the method-of-characteristics (MOC). In the solution process, three functions occur for each h_i

$$g_i^{(a)}\left(u_1^{(a)},u_2^{(a)}\right), \qquad g_i^{(b)}\left(u_1^{(b)},u_2^{(b)}\right), \qquad g_i^{(c)}\left(u_1^{(c)},u_2^{(c)}\right), \qquad i = 1, 2, 3 \quad \text{(K.6)}$$

where the gs are arbitrary functions of their arguments, and the u_1,u_2 equal the integration constants of the ordinary differential equations (ODEs). The gs are chosen such that

$$h_2 = h_2^{(a)} = h_2^{(b)} = h_2^{(c)}, \qquad i = 1, 2, 3 \tag{K.7}$$

(As will be demonstrated, it is this last step that breaks down.) Once an $h_i(x_j)$ function is established, the corresponding ξ_i coordinate is found by sequentially integrating the $\partial\xi_i/\partial x_j, j = 1, 2, 3$, derivatives in Equation (K.3).

The foregoing procedure is simplified by using the elliptic paraboloid (EP) model and Appendix J. For instance, the three PDEs for h_2 have the form

$$\frac{x_2}{r_2}\frac{\partial q_2^{(a)}}{\partial x_1} + \frac{\partial q_2^{(a)}}{\partial x_2} + 0\frac{\partial q_2^{(a)}}{\partial x_3} + \frac{x_2}{r_2^2}\frac{1}{1+\psi} = 0 \tag{K.8a}$$

$$\frac{\sigma x_3}{r_3}\frac{\partial q_2^{(b)}}{\partial x_1} + 0\frac{\partial q_2^{(b)}}{\partial x_2} + \frac{\partial q_2^{(b)}}{\partial x_3} + \frac{\sigma x_3}{r_3^2}\frac{1}{1+\psi} = 0 \tag{K.8b}$$

$$0 \frac{\partial q_2^{(c)}}{\partial x_1} + \frac{\sigma x_3}{r_3} \frac{\partial q_2^{(c)}}{\partial x_2} - \frac{x_2}{r_2} \frac{\partial q_2^{(c)}}{\partial x_3} + \left(\frac{1}{r_2} - \frac{1}{r_3} \right) \frac{\sigma x_2 x_3}{r_2 r_3} \frac{1}{1+\psi} = 0 \qquad \text{(K.8c)}$$

These have the following solutions:

$$h_2^{(a)} = \frac{1}{(1+\psi)^{1/2}} \, g_2^{(a)} \left(x_3, x_1 - \frac{x_2^2}{2r_2} \right) \qquad \text{(K.9a)}$$

$$h_2^{(b)} = \frac{1}{(1+\psi)^{1/2}} \, g_2^{(b)} \left(x_2, x_1 - \frac{\sigma x_3^2}{2r_3} \right) \qquad \text{(K.9b)}$$

$$h_2^{(c)} = \frac{1}{(1+\psi)^{1/2}} \, g_2^{(c)} \left(x_1, \frac{x_2^2}{2r_2} + \frac{\sigma x_3^2}{2r_3} \right) \qquad \text{(K.9c)}$$

Set

$$g_2^{(a)} = g_2^{(b)} = g_2^{(c)} = 1 \qquad \text{(K.10)}$$

to obtain

$$h_2 = \frac{1}{(1+\psi)^{1/2}} \qquad \text{(K.11a)}$$

With this h_2 value, integration of the $\partial \xi_2 / \partial x_j$ equations results in

$$\xi_2 = F = x_1 - \frac{x_2^2}{2r_2} - \frac{\sigma x_3^2}{2r_3} \qquad \text{(K.11b)}$$

in accord with Chapter 3.

For $q_1 = \ln h_1$, the EP model PDEs are

$$\frac{x_2}{r_2} \frac{\partial q_1^{(a)}}{\partial x_1} - \psi \frac{\partial q_1^{(a)}}{\partial x_2} + 0 \frac{\partial q_1^{(a)}}{\partial x_3} + \frac{x_2}{r_2^2} \frac{1}{1+\psi} = 0 \qquad \text{(K.12a)}$$

$$\frac{\sigma x_3}{r_3} \frac{\partial q_1^{(b)}}{\partial x_1} + 0 \frac{\partial q_1^{(b)}}{\partial x_2} - \psi \frac{\partial q_1^{(b)}}{\partial x_3} + \frac{\sigma x_3}{r_3^2} \frac{1}{1+\psi} = 0 \qquad \text{(K.12b)}$$

$$0 \frac{\partial q_1^{(c)}}{\partial x_1} + \frac{\sigma x_3}{r_3} \frac{\partial q_1^{(c)}}{\partial x_2} - \frac{x_2}{r_2} \frac{\partial q_1^{(c)}}{\partial x_3} + \left(\frac{1}{r_2} - \frac{1}{r_3} \right) \frac{\sigma x_2 x_3}{r_2 r_3} \frac{1+2\psi}{\psi(1+\psi)} = 0 \quad \text{(K.12c)}$$

Note the $2 \leftrightarrow 3$ symmetry between Equations (K.12a) and (K.12b). The three solutions are

$$h_1^{(a)} = \left(\frac{\psi}{1+\psi} \right)^{1/2} g_1^{(a)} \left(x_3, x_1 + \frac{r_2}{2} \ln \psi \right) \qquad \text{(K.13a)}$$

$$h_1^{(b)} = \left(\frac{\psi}{1+\psi}\right)^{\sigma/2} g_1^{(b)}\left(x_2, x_1 + \frac{\sigma r_3}{2} \ln \psi\right) \tag{K.13b}$$

$$h_1^{(c)} = \frac{1}{\psi^{1/2}(1+\psi)^{1/2}} g_1^{(c)}\left(x_1, \frac{x_2^2}{2r_2} + \frac{\sigma x_3^2}{2r_3}\right) \tag{K.13c}$$

Condition (K.7) results in

$$g_1^{(a)}\left(x_3, x_1 + \frac{r_2}{2}\ln\psi\right) = \left(\frac{1+\psi}{\psi}\right)^{(1-\sigma)/2} g_1^{(b)}\left(x_2, x_1 + \frac{\sigma r_3}{2}\ln\psi\right)$$

$$= \frac{1}{\psi} g_1^{(c)}\left(x_1, \frac{x_2^2}{2r_2} + \frac{\sigma x_3^2}{2r_3}\right) \tag{K.14}$$

When $\sigma = 0$, we can choose

$$g_1^{(a)} = 1, \qquad g_1^{(b)} = \frac{\dfrac{x_2}{r_2}}{\left(1 + \dfrac{x_2^2}{r_2^2}\right)^{1/2}}, \qquad g_1^{(c)} = \frac{2}{r_2}\left(\frac{x_2^2}{2r_2}\right) = \psi$$

with the result

$$h_1 = \left(\frac{\psi}{1+\psi}\right)^{1/2}$$

when $\sigma = 1$, $r = r_2 = r_3$, we can choose

$$g_1^{(a)} = g_1^{(b)} = 1, \qquad g_1^{(c)} = \frac{2}{r}\left(\frac{x_2^2 + x_3^2}{2r}\right) = \psi$$

with the same h_1 result. However, when $\sigma = 1$, $r_2 \neq r_3$, Equation (K.14) does not have a nontrivial solution.

For completeness, the result for h_3 and ξ_3 is provided. Because

$$\frac{\partial \xi_3}{\partial x_1} = 0 \tag{K.15}$$

only one PDE

$$\frac{\partial^2 \xi_3}{\partial x_3 \partial x_2} = \frac{\partial^2 \xi_3}{\partial x_2 \partial x_3} \tag{K.16}$$

is relevant. This becomes

$$\frac{x_2}{r_2}\frac{\partial q_3}{\partial x_2} + \frac{\sigma x_3}{r_3}\frac{\partial q_3}{\partial x_3} - \frac{1}{r_2 r_3}\left(\frac{x_2^2}{r_2} + \frac{\sigma x_3^2}{r_3}\right)\frac{1}{\psi} = 0 \qquad (K.17)$$

whose general solution is

$$h_3 = \frac{1}{\psi^{1/2}}\left(\frac{x_2}{r_2}\right)^{r_3/(r_3-r_2)}\left(\frac{\sigma x_3}{r_3}\right)^{-r_2/(r_3-r_2)} g_3\left[x_1, \left(\frac{x_2}{r_2}\right)^{r_2}\left(\frac{\sigma x_3}{r_3}\right)^{-r_3}\right] \qquad (K.18a)$$

With $g_3 = 1$, we have

$$h_3 = \frac{1}{\psi^{1/2}}\left(\frac{x_2}{r_2}\right)^{r_3/(r_3-r_2)}\left(\frac{\sigma x_3}{r_3}\right)^{-r_2/(r_3-r_2)} \qquad (K.18b)$$

The corresponding coordinate is

$$\xi_3 = (r_3 - r_2)\left(\frac{x_2}{r_2}\right)^{r_2/(r_2-r_3)}\left(\frac{\sigma x_3}{r_3}\right)^{-r_3/(r_3-r_2)} \qquad (K.19)$$

Note the $r_3 \neq r_2$ requirement.

With ξ_2 and ξ_3 known, the orthogonality condition can be used:

$$\nabla\xi_1 \cdot \nabla\xi_2 = 0, \quad \nabla\xi_3 \cdot \nabla\xi_1 = 0 \qquad (K.20)$$

to try to circumvent the requirement of first obtaining h_1. We thereby obtain

$$\frac{\partial \xi_1}{\partial x_1} - \frac{x_2}{r_2}\frac{\partial \xi_1}{\partial x_2} - \frac{\sigma x_3}{r_3}\frac{\partial \xi_1}{\partial x_3} = 0 \qquad (K.21a)$$

$$0\frac{\partial \xi_1}{\partial x_1} + \frac{r_2}{x_2}\frac{\partial \xi_1}{\partial x_2} - \frac{r_3}{\sigma x_3}\frac{\partial \xi_1}{\partial x_3} = 0 \qquad (K.21b)$$

These yield

$$\xi_1^{(a)} = g^{(a)}\left(x_1 + r_2 ln x_2, \; x_1 + \sigma r_3 ln x_3\right) \qquad (K.22a)$$

$$\xi_1^{(b)} = g^{(b)}\left(x_1, \; \frac{x_2^2}{2r_2} + \frac{\sigma x_3^2}{2r_3}\right) \qquad (K.22b)$$

and, again, there is no joint solution. As a check, one can show that

$$\nabla\xi_2 \cdot \nabla\xi_3 = 0$$

is satisfied.

We next demonstrate, again by example, that the existence difficulty is not unique to an EP shock or to h_1 and ξ_1. For this, an elliptic cone shock

$$F = \frac{x_1^2}{2r_1} - \frac{x_2^2}{2r_2} - \frac{x_3^2}{2r_3} = 0 \tag{K.23}$$

with a uniform freestream is utilized. The three PDEs for h_2 are

$$\frac{x_2}{r_2}\frac{\partial q_2^{(a)}}{\partial x_1} + \frac{x_1}{r_1}\frac{\partial q_2^{(b)}}{\partial x_2} + 0\frac{\partial q_2^{(a)}}{\partial x_3} + \left(\frac{1}{r_1} + \frac{1}{r_2}\right)\frac{x_1 x_2}{r_1 r_2}\frac{1}{|\nabla F|^2} = 0 \tag{K.24a}$$

$$\frac{x_3}{r_3}\frac{\partial q_2^{(b)}}{\partial x_1} + 0\frac{\partial q_2^{(b)}}{\partial x_2} + \frac{x_1}{r_1}\frac{\partial q_2}{\partial x_3} + \left(\frac{1}{r_1} + \frac{1}{r_2}\right)\frac{x_1 x_3}{r_1 r_3}\frac{1}{|\nabla F|^2} = 0 \tag{K.24b}$$

$$0\frac{\partial q_2^{(c)}}{\partial x_1} + \frac{x_3}{r_3}\frac{\partial q_2^{(c)}}{\partial x_2} - \frac{x_2}{r_2}\frac{\partial q_2^{(c)}}{\partial x_3} + \left(\frac{1}{r_2} - \frac{1}{r_3}\right)\frac{x_2 x_3}{r_2 r_3}\frac{1}{|\nabla F|^2} = 0 \tag{K.24c}$$

These have the respective solutions:

$$h_2^{(a)} = |\nabla F| \, g_2^{(a)}\left(x_3, \frac{x_1^2}{2r_1} - \frac{x_2^2}{2r_2}\right) \tag{K.25a}$$

$$h_2^{(b)} = \frac{1}{|\nabla F|} \, g_2^{(b)}\left(x_2, \frac{x_1^2}{2r_1} - \frac{x_3^2}{2r_3}\right) \tag{K.25b}$$

$$h_2^{(c)} = |\nabla F| \, g_1^{(c)}\left(x_1, \frac{x_2^2}{2r_2} + \frac{x_3^2}{2r_3}\right) \tag{K.25c}$$

In this case, h_2 does not exist if $r_2 \neq r_3$. For a circular cone, $r_2 = r_3$, a solution does exist—that is,

$$h_2 = \frac{1}{|\nabla F|} \tag{K.26}$$

As shown in Chapter 3, an explicit solution for the ξ_i exists when the flow is two-dimensional or axisymmetric. Because an orthonormal, flow-plane-based basis, at every shock point, is readily established, a corresponding three-dimensional coordinate system was expected to exist. In the general case, however, this does not occur even when the freestream is uniform. Two arbitrarily positioned, closely spaced shock points cannot be represented by a single coordinate system. A local solution in which the ξ_i are linearly related to the $\bar{x}_j$ can be determined but was also unsuccessful.

Appendix L: Unsteady State 2 Parameters

The basic data required for a solution is

$$\gamma, \quad F(x_i,t), \quad v'_{1,j}(x_i,t), \quad p_1(x_i,t), \quad \rho_1(x_i,t)$$

From these, we first evaluate:

$$F_{x_i}, \quad F_t, \quad F_{tt}, \quad F_{x_ix_j}, \quad F_{x_it}, \quad v_{1,jt}, \quad p_{1t}, \quad \rho_{1t} \tag{L.1}$$

The following is then evaluated:

$$|\nabla F| = \left(\sum F_{x_i}^2\right)^{1/2} \tag{L.2}$$

$$v_{1,i} = v'_{1,i} + \frac{F_t F_{x_i}}{|\nabla F|^2}, \quad i = 1,2,3 \tag{L.3}$$

$$V_1 = \left(\sum v_{1,i}'^2 + \frac{2F_t}{|\nabla F|^2}\sum F_{x_i}v'_{1,i} + \frac{F_t^2}{|\nabla F|^2}\right)^{1/2} \tag{L.4}$$

$$v_{1,ii} = v'_{1,it} + \frac{F_{tt}F_{x_i}}{|\nabla F|^2} + \frac{F_t F_{x_it}}{|\nabla F|^2} - \frac{2F_t F_{x_i}}{|\nabla F|^4}\sum F_{x_j}F_{x_jt}, \quad i = 1,2,3 \tag{L.5}$$

$$M_1^2 = \frac{V_1^2\rho_1}{\gamma p_1} \tag{L.6}$$

$$\frac{M_{1t}}{M_1} = \frac{1}{V_1^2}\sum v_{1,i}v_{1,it} - \frac{1}{2}\frac{p_{1t}}{p_1} + \frac{1}{2}\frac{\rho_{1t}}{\rho_1} \tag{L.7}$$

$$\sin\beta = \frac{1}{V_1|\nabla F|}\sum v_{1,i}F_{x_i} \tag{L.8}$$

$$\beta_t = \frac{\chi}{V_1^2|\nabla F|^2}\left\{|\nabla F|^2\left[V_1^2\sum v_{1,it}F_{x_i} - \left(\sum v_{1,i}F_{x_j}\right)\sum v_{1,j}v_{1,jt}\right]\right.$$

$$\left. + V_1^2\left[|\nabla F|^2\sum v_{1,i}F_{x_it} - \left(\sum v_{1,i}F_{x_i}\right)\sum F_{x_j}F_{x_jt}\right]\right\} \tag{L.9}$$

$$\chi = \frac{1}{V_1 |\nabla F| \cos\beta} \tag{L.10}$$

$$\frac{\chi_t}{\chi} = -\frac{1}{V_1^2} \sum v_{1,i} v_{1,it} - \frac{1}{|\nabla F|^2} \sum F_{x_i} F_{x_i t} + \tan\beta \beta_t \tag{L.11}$$

$$w = M_1^2 \sin^2\beta \tag{L.12}$$

$$w_t = 2w \left(\frac{M_{1t}}{M_1} + \cot\beta \, \beta_t \right) \tag{L.13}$$

$$A = \frac{\gamma+1}{2} \frac{M_1^2 \sin\beta \cos\beta}{1 + \frac{\gamma-1}{2} w}, \quad B = 1 + A^2 \tag{L.14}$$

$$A_t = \frac{(\gamma+1) M_1^2}{\left(1 + \frac{\gamma-1}{2} w\right)^2} \left\{ 2 \frac{M_{1t}}{M_1} \sin\beta \cos\beta + \left[\cos^2\beta - \frac{1}{2}\left(1 + \frac{\gamma-1}{2} M_1^2\right)\sin^2\beta \right]\beta_t \right\} \tag{L.15}$$

$$K_1 = v_{1,3} F_{x2} - v_{1,2} F_{x3}, \quad K_2 = v_{1,1} F_{x3} - v_{1,3} F_{x1}, \quad K_3 = v_{1,2} F_{x1} - v_{1,1} F_{x2} \tag{L.16}$$

$$L_i = |\nabla F|^2 v_{1,i} - \left(\sum v_{j,1} F_{x_j} \right) F_{x_i} \tag{L.17}$$

$$L_{it} = |\nabla F|^2 v_{1,it} - F_{x_i} \sum F_{x_j} v_{1,jt} - F_{x_i t} \sum v_{1,j} F_{x_j} + 2\left(\sum F_{x_j} F_{x_j t} \right) v_{1,i} - F_{x_i} \sum v_{1,j} F_{x_j t} \tag{L.18}$$

$$p_2 = \frac{2}{\gamma+1} p_1 \left(\gamma w - \frac{\gamma-1}{2} \right) \tag{L.19}$$

$$p_{2t} = \frac{2}{\gamma+1} \left[\left(\gamma w - \frac{\gamma-1}{2} \right) p_{1t} + 2\gamma \, p_1 w \left(\frac{M_{1t}}{M_1} + \cot\beta \beta_t \right) \right] \tag{L.20}$$

$$\rho_2 = \frac{\gamma+1}{2} \rho_1 \frac{w}{1 + \frac{\gamma-1}{2} w} \tag{L.21}$$

$$\rho_{2t} = \frac{\gamma+1}{2} \frac{w}{\left(1 + \frac{\gamma-1}{2} w\right)^2} \left[\left(1 + \frac{\gamma-1}{2} w\right) \rho_{1t} + 2\rho_1 \left(\frac{M_{1t}}{M_1} + \cot\beta \beta_t \right) \right] \tag{L.22}$$

$$u_2 = V_1 \cos\beta \tag{L.23}$$

$$v_2 = \frac{2}{\gamma + 1} V_1 \frac{1 + \frac{\gamma - 1}{2} w}{M_1^2 \sin \beta} \tag{L.24}$$

Equations for u_{2t} and v_{2t} are not provided, since they do not appear in Equation (9.105). For the s and b tangential derivatives, utilize Appendix H.4.

$$V_2 = \frac{2}{\gamma + 1} V_1 \frac{XB^{1/2} \sin \beta}{w} \tag{L.25}$$

$$\vec{V}_2 = \sum v_{2,i} \hat{1}_i = \frac{V_2}{B^{1/2}} (\hat{n} + A\hat{t}) \tag{L.26}$$

$$V_{2,i} = \frac{V_2}{|\nabla F| B^{1/2}} (F_{x_i} + \chi AL_i) \tag{L.27}$$

$$J = \frac{1}{V_1^2} \sum v_{1,j} v_{1,jt} - \frac{1}{|\nabla F|^2} \sum F_{x_j} F_{xjt} - \frac{2}{1 + \frac{\gamma - 1}{2} w} \left[\frac{M_{1t}}{M_1} + \frac{1}{2} \left(1 - \frac{\gamma - 1}{2} w \right) \cot \beta \beta_t \right] \tag{L.28}$$

$$H_i = \frac{1}{F_{x_i} + \chi AL_i} \left(F_{xit} + AL_i \chi_t + \chi L_i A_t + \chi AL_{it} \right) \tag{L.29}$$

$$v_{2,it} = v_{2,i} (J + H_i) \tag{L.30}$$

Problems

1. Start with Equation (2.27) and derive Equation (2.28).
2. The thermodynamics of a van der Waals gas is based on the thermal equation of state

$$p = \frac{\rho R T}{1 - \beta \rho} - \alpha \rho^2$$

where α, β, and R are constants. By introducing reduced variables

$$p = p_c p_r, \qquad \rho = \rho_c \rho_r, \qquad T = T_c T_r$$

where a c subscript denotes a critical point value

$$p_c = \frac{\alpha}{27 \beta^2}, \qquad \rho_c = \frac{1}{3\beta}, \qquad T_c = \frac{8\alpha}{27 \beta R}$$

we obtain

$$p_r = \frac{8 \rho_r T_r}{3 - \rho_r} - 3 \rho_r^2$$

For this equation of state, the constant volume specific heat c_v is a function only of the temperature. For purposes of simplicity, assume c_v to be a constant. As a consequence, the reduced enthalpy, entropy, and speed of sound are

$$h_r = \frac{h - h_C}{R T_c} = \frac{c_v}{R}(T_r - 1) + \frac{3 T_r}{3 - \rho_r} + \frac{9}{4}(1 - \rho_r) - \frac{3}{2}$$

$$s_r = \frac{s - s_c}{R} = \ln\left(\frac{4 - \rho_r}{2\rho_r}\right) + \frac{c_v}{R} \ln T_r$$

$$a_r^2 = \frac{a^2}{a_c^2} = \frac{a^2}{\dfrac{2\,R\,\alpha}{3\,c_v\,\beta}} = \left(1 + \frac{c_v}{R}\right)\frac{4 T_r}{(3 - \rho_r)^2} - \frac{c_v}{R}\rho_r$$

For this gas, determine the counterpart to Equation (2.28) in the form

$$F(\theta, \beta; m, c_v/R, p_{r1}, \rho_{r1}) = 0$$

where

$$m^2 = \frac{V_1^{*2}}{RT_c}$$

3. With Equation (3.12), show that Equation (3.21) is identically satisfied for $j = 2$.
4. Utilize Equation (6.21), with $y = R$, to obtain the scale factors and coordinate transformation for this surface. Your answer should be in terms of r, β_∞, z, x_i, and R. Simplify your results as much as possible. (Hint: Use Equations 3.28 and 3.29.)
5. A two-dimensional or axisymmetric shock has the shape

$$2rx_1 - x_2^2 - \sigma x_3^2 = 0$$

where r is the radius of curvature at the nose of the shock, and the upstream flow is uniform. Determine $\hat{t}$, $\hat{n}$, $\hat{b}$, the h_i, and the ξ_i in terms mostly of the x_i, R, and β.
6. Start with the $\tan\theta$ equation in Appendix E.1 and derive the subsequent equation for $\sin\theta$.
7. Assume a steady, three-dimensional flow of perfect gas. Use

$$\vec{V} = u\hat{t} + v\hat{n} + \bar{w}\hat{b}$$

$$\nabla = \hat{t}\,\frac{\partial}{\partial s} + \hat{n}\,\frac{\partial}{\partial n} + \hat{b}\,\frac{\partial}{\partial b}$$

and evaluate the following at state 2:
   (a) $D\vec{V}/Dt$.
   (b)   Use the part (a) general result to evaluate the three scalar Euler momentum equations for a two-dimensional or axisymmetric shock. Do not, at this time, delete any terms.
   (c)   Utilize Appendix E to evaluate $(\partial p/\partial s)_2$ and compare your result with the $(\partial p/\partial s)_2$ in Appendix E. What changes, if any, are needed for agreement?
   (d)   Repeat part (c) for $(\partial p/\partial n)_2$.
8. Show that the entropy satisfies

$$\left(\frac{\partial S}{\partial \tilde{s}}\right)_2 = 0$$

where the entropy is given by Equation (6.4). Utilize Equation (5.10a), but do not use the relation between the entropy and the stagnation pressure.

9. Let $\hat{e}_{no}$ be a unit vector in the flow plane that is normal to $\vec{V}_2$, and is for a right-handed system with $\vec{V}_2$ and $\hat{b}$. Use Appendix E to show that

$$\hat{e}_{no} = \frac{1}{B^{1/2}}\left(\hat{t} - A\hat{n}\right)$$

10. Derive some of the $mg_3 + g_4, \ldots, (\partial T/\partial n)_2$ equations listed in Section 5.1.

11. A normal shock has a speed $V_s(t)$. The uniform and steady upstream flow has a constant velocity, $V_1\hat{i}_1$, that is perpendicular to the shock. Develop equations for $(\partial p/\partial n)_2$, $(\partial \rho/\partial n)_2$, and $(\partial V^*/\partial n)_2$, where n is measured from the shock and is positive in the downstream direction. You will need the unsteady, one-dimensional Euler equations. These should be transformed from (x,t) coordinates to (n,τ) coordinates, where $\tau = t$ and $n = F(x,t)$, and $F = 0$ at the shock's location. Do not assume a specific form for the enthalpy (i.e., use Equation 2.26). This means Appendix E, which requires a perfect gas, is not used. Results will depend on time derivatives, such as $(\partial p/\partial \tau)_2$ and dV_s/dt. It is beyond the scope of this problem to evaluate $(\partial p/\partial \tau)_2$, for a perfect gas, using Appendix E.1, where p_1 is a constant. However, w equals M_1^{*2}, where this Mach number is based on V_1^*.

12. A spherically symmetric flow is caused by an intense point explosion in a uniform atmosphere. Assume air to be a perfect gas and the Euler equations in unsteady, spherical coordinates are

$$\rho_t + V\rho_r + \rho\left(V_r + \frac{2V}{r}\right) = 0$$

$$V_t + VV_r + \frac{1}{\rho}p_r = 0$$

$$h_{ot} + Vh_{or} - \frac{1}{\rho}p_t = 0$$

where the stagnation enthalpy is

$$h_o = \frac{\gamma}{\gamma - 1}\frac{p}{\rho} + \frac{1}{2}V^2$$

Eliminate h_o in favor of p, ρ, and V. Use a coordinate system fixed at the shock by introducing

$$R(r,t) = r_s(t) - r, \quad \tau = t, \quad W = V_s - V$$

where r_s is the radius of the spherical shock and the shock speed, V_s, is dr_s/dt.

(a)   Determine the Euler equations in terms of these variables.

(b)   Define the shock Mach number $M_s = (V_s/a_1)$, where a_1 is the speed of sound ahead of the shock, and develop jump conditions for p, ρ, and W assuming a strong shock (i.e., $M_s \gg 1$).

(c)   With this assumption, determine $\rho_{2R}[=(\partial\rho/\partial R)_2]$, p_{2R}, and W_{2R} in terms of r_s and its derivatives just behind the shock. Obtain simplified results for these derivatives when $\gamma = 1.4$ and $r_s = ct^{2/5}$, where c is a constant.

13. Consider a steady, two-dimensional or axisymmetric flow of a perfect gas. Let $\tilde{s}$ and $\tilde{n}$ be intrinsic coordinates where $\tilde{s}$ and $\tilde{n}$ are along and normal to the streamlines.

(a)   Derive an equation for $(\partial p/\partial\tilde{n})_2$ that has the form of the Appendix E.3 equations.

(b)   The curvature of the shock κ_s and the curvature of a streamline κ_o, just downstream of the shock, are given by

$$\kappa_s = -\beta', \quad \kappa_o = -\left(\frac{\partial\theta}{\partial\tilde{s}}\right)_2$$

The minus signs mean that both curvatures are positive when the shock appears as shown in Figure 5.3. Write a relation between the curvatures.

(c)   Evaluate the curvatures when

$$\beta = 90°, \quad \beta' = -\frac{1}{R_s}$$

14. The Crocco point is defined by

$$\left(\frac{\partial\theta}{\partial\tilde{s}}\right)_2 = 0$$

This derivative can be written as

$$\left(\frac{\partial\theta}{\partial\tilde{s}}\right)_2 = C_1\left(G_{cp}\beta' + \sigma C_2\right)$$

where cp stands for Crocco point and

$$C_1 = \frac{2}{\gamma+1}\frac{A}{X^2 Z B^{3/2}}$$

$$G_{cp} = \frac{\gamma+1}{4}(mg_5 + g_6) - 2X^2 Z$$

$$C_2 = XYZ \frac{\cos\beta}{y}$$

In the two-dimensional case, the Crocco point is determined by $G_{cp} = 0$ or $\beta' = 0$, hereafter excluded. With γ and M_1 fixed, $G_{cp} = 0$ is a cubic equation for $\sin^2\beta_{cp}$. Develop a computer code to determine β_{cp} for (eight cases)

$$\gamma = 1.4, \ 1.6667$$

$$M_1 = 2, \ 3, \ 4, \ 6$$

Compare β_{cp} with β^* (Equation 5.14) and with the detachment wave angle, β_d, given by Equation (7.10). If there are no bugs, these β_{cp} values satisfy

$$\beta^* < \beta_{cp} < \beta_d$$

Tabulate all three β values.

15. Develop a code to compute $d(\mu - \theta)/d\beta$, Equation (5.13b), as a function of γ, M_1, and β. Tabulate the derivative versus β for several M_1 values using $\gamma = 1.4$.
16. Use Appendix E to evaluate

$$\frac{dp}{d\eta}, \qquad \frac{d\rho}{d\eta}, \qquad \frac{dM^2}{d\eta}$$

just downstream of the conical shock in a Taylor-Maccoll flow (see Section 5.5).
17. Solve.
   (a)   A conical shock has

$$\gamma = 1.4, \quad M_1 = 3, \quad \beta = 30°$$

where $\theta_b = 20.5°$ (NACA 1135). Determine M_2, θ_2, μ_2, and β^*. Evaluate

$$\frac{1}{p_1}\left(\frac{\partial p}{\partial n}\right)_2, \qquad \left(\frac{\partial M^2}{\partial n}\right)_2, \qquad \left(\frac{\partial \theta}{\partial \tilde{s}}\right)_2$$

where each answer is a constant divided by y. Evaluate

$$\frac{1}{p_1}\left(\frac{\partial p}{\zeta_o}\right)_2, \qquad \frac{1}{p_1}\left(\frac{\partial p}{\zeta_\pm}\right)_2$$

Sketch the ζ_o, $\zeta_\pm$ curves and the shock with respect to x.

(b)   Repeat part (a) for the lower half of an inverted conical shock where $\beta = 210°$.

18. A normal detonation wave has cellular structure. Consider such a wave in an air/hydrogen mixture in which a given cell has an approximately spherical shock with a 3 mm radius. Ignore the effect of the downstream combustion process. Let θ be the angle measured from the center of the sphere. Upstream of the shock the temperature is 300 K and the gas constant is 450 J/kg-K. Tabulate the dimensional vorticity, $\bar{\omega}_2$, when $\theta = 0, 10, 20°$ and $M_1 = 3, 6$.

19. With the assumptions used in Chapter 3, Emanuel (1986, p. 270), shows that the vorticity can be written as

$$\vec{\omega} = \left(\frac{1}{h_1} \frac{\partial v_2}{\partial \xi_1} - \frac{1}{h_2} \frac{\partial v_1}{\partial \xi_2} + v_2 \kappa_2 - v_1 \kappa_1 \right) \hat{e}_3$$

(a)   Derive a form for $\vec{\omega}_2$ consistent with Appendix E.1.
(b)   Use the result of part (a) to derive a form for $\vec{\omega}_2/(\beta' V_1)$ that depends only on the θ and β angles (Truesdell, 1952).

20. Solve.
(a)   Use v_2 and $(\partial v/\partial n)_2$ to estimate the nondimensional shock wave stand-off distance $\Delta_{est}(= \bar{\Delta}_{est}/\bar{R}_b)$ for a convex shock. This estimate assumes a linear variation along the stagnation streamline for v. See Figure 6.1 and you will need r given by Equation (6.19b).
(b)   With $\gamma = 1.4$, compute Δ_{est} when

$$M_1 = 1, 2, 4, 6$$

for both two-dimensional and axisymmetric flows, and compare with the Δ given by Equation (6.19a).
(c)   Discuss your results.

21. Conditions for a triple point are

$$\gamma = 1.4, \quad M_1 = 3, \quad \beta_I = 46.04°, \quad \beta_R = 55.17°, \quad \text{type (b)}$$

(a)   Determine $M_2, M_3, M_4,$ and $\bar{\beta}_I, \bar{\beta}_R, \bar{\beta}_M,$ and $\bar{\theta}_{SS}$.
(b)   Is M weak or strong? Is R inverted and weak or strong?

22. Evaluate the vorticity, ω_2, for the source flow model when Equations (8.61) apply. Normalize ω_2 with V_1 and any reference length. What does the sign of ω_2 indicate and why?

23. Use Equation (G.17a,b) to analytically evaluate $(\partial p/\partial s)_1/p_1$ and $(\partial p/\partial b)_1/p_1$ for the flow model in Section 8.7. Then use Equation (8.61) for a numerical evaluation of both derivatives and compare with the $p_{1s}[= (\partial p/\partial s)_1/p_1]$ value in Equation (8.62c). (Hint: Consider γ, β', φ', and M_{1na} as given, and first evaluate F and V_1 in terms of $x_1, x_2 (= x, y)$ and M_1.)

24. Determine $z, y, \beta,$ and β' values for the Thomas points when

$$\gamma = 1.4, \quad M_1 = 2, \quad \sigma = 0, 1, \quad \theta_b = 15°$$

Use Equation (6.21) for the shock shape and Table 6.1. Each σ value has one Thomas point. Summarize your answers at the end.

25. Use $\gamma = 1.4$ and Equations (7.10) and (5.20) to evaluate the sign of $(1/p_1)$ $(\partial p/\partial \tilde{s})_2$ at the detachment point of a convex, two-dimensional, detached shock for M_1 values ranging from 1.2 to 6. What does this tell you about the location of the Thomas point?

26. Solve:

(a) Evaluate the parameters listed in Appendix J.5 when

$$\gamma = 1.4, \quad M_1 = 3, \quad w = 4, \quad r_2 = r_3 = 2, \quad \sigma = 0, 1$$

(b) Compare the part (a) answers for the s and n derivatives of u, v, p, and ρ with results from Appendix E.

References

Ames Research Staff, Equations, Tables, and Charts for Compressible Flow, *NACA Report* 1135, 1953.

Azevedo, D. J., and Liu, C. S., Engineering Approach to the Prediction of Shock Patterns in Bounded High-Speed Flows, *AIAA J.* 31, 83 (1993).

Ben-Dor, G., Igra, O., and Elperin, T., editors, *Handbook of Shock Waves*, Vols. 1, 2, 3, Academic Press, New York (2001).

Ben-Dor, G., *Shock Wave Reflection Phenomena*, 2nd ed., Springer, New York (2007).

Ben-Dor, G., and Glass, I., Nonstationary Oblique Shock-Wave Reflections: Actual Isopycnics and Numerical Experiments, *AIAA J.* 16, 1146–1153 (1978).

Billig, F. S., Shock-Wave Shapes around Spherical- and Cylindrical-Nosed Bodies, *J. Spacecraft and Rockets* 4, 822 (1967).

Borovoy, V. Ya., Chinilov, A. Yu., Gusev, V. N., Struminskaya, I. V., Délery, J., and Chanetz, B., Interference between a Cylindrical Bow Shock and a Plane Oblique Shock, *AIAA J.* 35, 1721 (1997).

Courant, R., *Differential and Integral Calculus*, Vol. II, Interscience, New York, p. 125 (1949).

Courant, R., and Friedrichs, K. O., *Supersonic Flow and Shock Waves*, Interscience, New York, pp. 331–350 (1948).

Edney, B. E., Effects of Shock Impingement on the Heat Transfer around Blunt Bodies, *AIAA J.* 6, 15 (1968).

Emanuel, G., *Gasdynamics Theory and Applications*, AIAA Education Series, Washington, DC (1986).

Emanuel, G., *Analytical Fluid Dynamics*, 2nd ed., CRC Press, Boca Raton, FL (2001).

Emanuel, G., Vorticity in Unsteady, Viscous, Reacting Flow Just Downstream of a Curved Shock, *AIAA J.* 45, 2097 (2007).

Emanuel, G., Vorticity Jump across a Shock in Nonuniform Flow, *Shock Waves* 21, 71–72 (2011).

Emanuel, G., and Hekiri, H., Vorticity and Its Rate of Change Just Downstream of a Curved Shock, *Shock Waves* 17, 85–94 (2007).

Emanuel, G., and Liu, M. -S., Shock Wave Derivatives, *Phys. Fluids* 31, 3625 (1988).

Emanuel, G., and Yi, T. H., Unsteady Oblique Shock Waves, *Shock Waves* 10, 113–117 (2000).

Ferri, A., Supersonic Flows with Shock Waves, in *General Theory of High-Speed Aerodynamics*, ed. W. R. Sears, High-Speed Aerodynamics and Jet Propulsion, Vol. VI, Princeton University Press, NJ, p. 677 (1954).

Glass, I. I., and Sislian, J. P., *Nonstationary Flows and Shock Waves*, Clarendon Press, UK (1994).

Goldstein, H., *Classical Mechanics*, Sect. 4-3, Addison-Wesley, Cambridge, MA (1950).

Gradshteyn, I. S., and Ryzhik, I. M., *Tables of Integrals, Series, and Products*, p. 277, Eq. (13), Academic Press, New York (1980).

Hayes, W. D., The Vorticity Jump across a Gasdynamic Discontinuity, *J. Fluid Mech.* 2, 595 (1957).

Hayes, W. D., and Probstein, R. F., *Hypersonic Flow Theory*, Section 6.5, Academic Press, New York (1959).

Hekiri, H., and Emanuel, G., Shock Wave Triple-Point Morphology, *Shock Waves* 21, 511–521 (2011).

Henderson, L. F., On the Confluence of Three Shock Waves in a Perfect Gas, *Aero. Quart.* 15, 181–197 (1964).

Henderson, L. R. F., and Menikoff, R., Triple-Shock Entropy Theorem and Its Consequences, *J. Fluid Mech.* 366, 179 (1998).

Hornung, H., Regular and Mach Reflection of Shock Waves, *Ann. Rev. Fluid Mech.* 18, 33 (1986).

Hornung, H., Deriving Features of Reacting Hypersonic Flow from Gradients at a Curved Shock, *AIAA J.* 48, 287 (2010).

Ivanov, M. S., Markelov, G. N., Kudryavtsev, A. N., and Gimelshein, S. F., Numerical Analysis of Shock Wave Reflection Transition in Steady Flows, *AIAA J.* 36, 2079 (1998).

Kalghatgi, G. T., and Hunt, B. L., The Three-Shock Confluence Problem for Normally Impinging Overexpanded Jets, *Aeron. Quar.* XXVI, 117 (1975).

Kaneshige, M. J., and Hornung, H. G., Gradients at a Curved Shock in Reacting Flow, *Shock Waves* 9, 219 (1999).

Kanwal, R. P., Determination of the Vorticity and the Gradient of Flow Parameters behind a Three-Dimensional Unsteady Curved Shock Wave, *Arch. Ration. Mech. Anal.*, 1, 225 (1958a).

Kanwal, R. P., On Curved Shock Waves in Three-Dimensional Gas Flows, *Quart. Appl. Math.* 16, 361 (1958b).

Lin, C. C., and Rubinov, S. I., On the Flow behind Curved Shocks, *J. Math. Phys.* 27, 105 (1948).

Milne-Thomson, L. M., Elliptic Integrals, in *Handbook of Mathematical Functions*, National Bureau of Standards, *Appl. Math. Ser.* 55 (1972).

Mölder, S., Internal, Axisymmetric, Conical Flow, *AIAA J.* 5, 1252 (1967).

Mölder, S., Curved Aerodynamic Shock Waves, Ph.D. thesis, Department of Mechanical Engineering, McGill University (2012).

Morse, P. M., and Feshbach, H., *Methods of Theoretical Physics*, McGraw-Hill, New York, Section 1.3 (1953).

Mouton, C. A., and Hornung, H. G., Mach Stem Height and Growth Rate Predictions, *AIAA J.* 45, 1977 (2007).

Rand, R. H., Computer Algebra in Applied Mathematics, an Introduction to MACSYMA, Research Notes in Mathematics, Vol. 94, Pitman, London (1984).

Serrin, J., Mathematical Principals of Classical Fluid Mechanics, in *Encyclopedia of Physics*, Vol. VIII/1, pp. 155–157, Springer, Berlin (1959).

Sterbenz, W. H., and Evvard, J. C., Criterions for Prediction and Control of Ram-Jet Flow Pulsations, NACA TN 3506 (August 1955).

Stoker, J. J., *Differential Geometry*, p. 392, John Wiley, New York (1969).

Thomas, T. Y., On Curved Shock Waves, *J. Math. Phys.* 26, 62 (1947).

Thomas, T. Y., Calculation of the Curvature of Attached Shock Waves, *J. Math. and Phys.* 27, 279 (1948).

Truesdell, C., On Curved Shocks in Steady Plane Flow of an Ideal Fluid, *J. Aeron. Sci.* 19, 826 (1952).

Uskov, V. N., and Chernyshov, M. V., Special and Extreme Triple Shock Wave Configurations, *J. Appl. Mech.* 47(4), 492 (2006).

Uskov, V. N., and Mostovykh, P. S., Interference of Stationary and Non-Stationary Shock Waves, *Shock Waves* 20, 119–129 (2010).

Van Dyke, M., *An Album of Fluid Motion*, Parabolic Press, Stanford, CA (1982).

Vincenti, W. G., and Kruger, C. H., Jr., *Introduction to Physical Gas Dynamics*, John Wiley, New York (1965).

Wilson, L. N., Inflections in Bow Shock Shape at Hypersonic Speeds, *AIAA J.* 5, 1532 (1967).

Yi, T. H., *Study of an Unsteady, Oblique Shock Wave*, MS Thesis, School of Aerospace and Mechanical Engineering, University of Oklahoma, Norman, OK (1999).

Yi, T. H., and Emanuel, G., Unsteady Shock Generated Vorticity, *Shock Waves* 10, 179–184 (2000).

Zel'dovich, Ya. B., and Raizer, Yu. P., *Physics of Shock Waves and High-Temperature Phenomena*, Vols. I and II, Academic Press, New York (1966).

Zucrow, M. J., and Hoffman, J. D., *Gas Dynamics*, Vol. I, Sections 7-5 and 7-8, John Wiley, New York (1976).

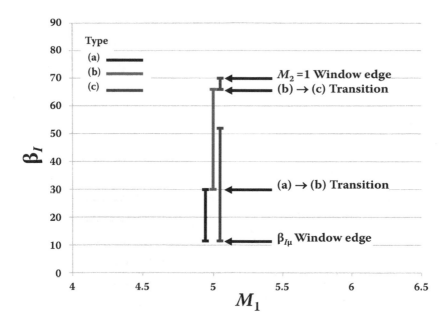

FIGURE 7.6 M_1, β_I sketch with labeling for $\gamma = 1.4$, $M_1 = 5$.

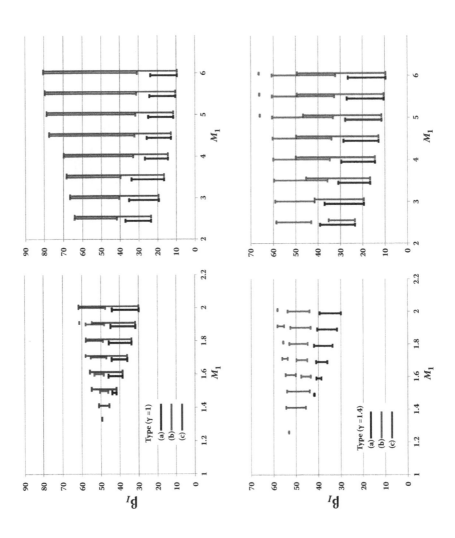

FIGURE 7.7 Solution types when M_1 equals its onset value, 1.4 (0.1) 2(0.5) 6, (a) $\gamma = 1$, (b) $\gamma = 1.4$.

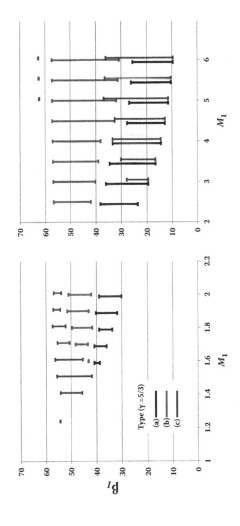

FIGURE 7.7 Solution types when M_1 equals its onset value, 1.4 (0.1) 2(0.5) 6, (c) $\gamma = 5/3$.

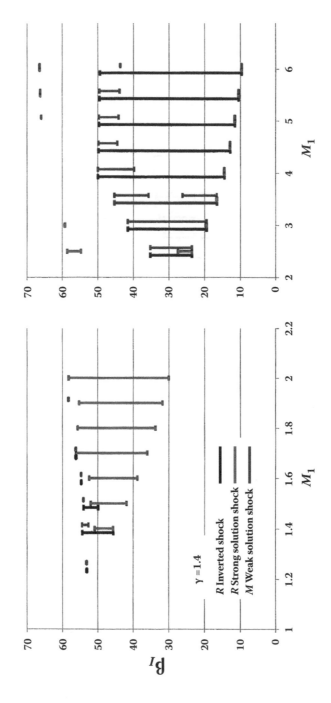

FIGURE 7.8 Graph indicating when R is inverted, R is a strong solution shock, and M is a weak solution shock. $\gamma = 1.4$.

Index